Math Dictionary

The Easy, Simple, Fun Guide to Help Math Phobics Become Math Lovers

Eula Ewing Monroe
Professor of Mathematics Education
Brigham Young University

Professor Emeritus
Western Kentucky University

BOYDS MILLS PRESS
HONESDALE, PENNSYLVANIA

Published by Boyds Mills Press, Inc.
A Highlights Company
815 Church Street
Honesdale, Pennsylvania 18431
www.boydsmillspress.com
Printed in China

First edition, 2006

Library of Congress Cataloging-in-Publication Data
Monroe, Eula Ewing.
   Math dictionary : the easy, simple, fun guide to help math phobics become math lovers/
Eula Ewing Monroe.—1st ed.
      p. cm
Includes bibliographical references.
ISBN 978-1-59078-413-6 (paperback : alk. paper)
   1. Mathematics—Dictionaries. I. Title

QA5.M575 2006
510'.3—dc22

2005050725

Book and cover design by Jeff George

10 9 8 7 6 5 4 3 2 1

Contents

Acknowledgments

My deepest appreciation to:

Robert Panchyshyn, professor emeritus, Western Kentucky University, a valued colleague and lifetime friend, for his leadership in analyzing vocabulary needed to learn mathematics and for his inspiration and encouragement for further study.

Wilburn Jones, professor emeritus, Western Kentucky University, my mentor in mathematics content, for his review of the original manuscript for accuracy and appropriateness for developing mathematicians.

The administration of the Brigham Young University School of Education and my colleagues in the Department of Teacher Education for their enthusiastic support of my professional endeavors.

Kent Brown Jr., publisher, Boyds Mills Press, for his keen understanding of the power of the printed word to enrich human lives, his steadfast good humor, and his leadership in publishing this work.

Susan B. Layne for her zest for language and for her lively interest in things mathematical.

Amanda, Becky, Bobbi, Brittany, Chris, Danae, Jessica, Jennifer, Julie, Megan, Melissa, two Michelles, Natalie, and Robin for their seemingly infinite patience in assisting with the details of my work.

Matt, Jamie, and Roger for always loving me unconditionally.

D'Lynn, Kathy, Diana, members of the Baptist Student Union, and the pastor and congregation of First Baptist Church of Provo for their continuing support and prayers.

My Lord and Savior for the sure knowledge that this work has been within His plan for my life and that all things are possible through Him.

Message to Teachers and Parents

What Is in This Dictionary?

Math Dictionary: The Easy, Simple, Fun Guide to Help Math Phobics Become Math Lovers is a handy reference for the language of mathematics. Students, parents, and teachers will find it helpful, as will anyone who is looking for clear and simple definitions of basic terms needed in the study of mathematics.

This dictionary is written in user-friendly and mathematically appropriate language; everyday words are used whenever possible in definitions and examples. When technical terms are needed for an accurate description of a mathematical concept, as so often is the case, they consist of words the reader may already know or can find readily using the *See also* feature within the dictionary. Examples, diagrams, pictures, and interesting facts are included to clarify definitions and provide bridges to other mathematical ideas and to the real world.

More than five hundred entries are included in this dictionary, carefully selected through extensive study of the terms found in mathematics textbooks. Most of the terms are core vocabulary needed for the further study of mathematics (for example, *multiple*). A few lesser-known terms (for example, *googol*, which is similar to the name of the widely known Internet search engine Google), are included because of the interesting mathematical ideas they represent.

Math lovers will enjoy using this reference, which is abundant with fascinating facts about how math relates to the natural world. And math phobics will find it a nonthreatening guide into the exciting field of mathematics. It may also help answer the perennial question, "What are we ever going to use math for, anyway?"

Why Learn the Language of Mathematics?

Benjamin Whorf, a noted linguist, hypothesized that language is necessary for higher-level thinking; moreover, an individual's language structure and development shape his or her understanding of the world (Carroll, 1956). Lev Semenovich Vygotsky, a noted Russian cognitive psychologist, theorized that the intellectual development of children is dramatically affected by their interaction with language (Vygotsky, 1962, 1978, cited in Reutzel & Cooter, 1996).

The work of these and other scholars who have studied relationships between language and thought supports the current emphasis on communication in the learning of mathematics.

In 1989 the National Council of Teachers of Mathematics (NCTM) published a landmark document entitled *Curriculum and Evaluation Standards for School Mathematics*, followed by *Professional Standards for Teaching Mathematics* (1991), *Assessment Standards for School Mathematics* (1995), and *Principles and Standards for School Mathematics* (2000). These documents articulate a near consensus among the mathematics education community regarding the need for rich and meaningful communication in the teaching and learning of mathematics.

The ability to communicate mathematically is viewed as a central goal for each learner, to be addressed in all aspects of mathematics instruction and assessment.

How to Use This Dictionary

Index Letters indicate the letter of the alphabet that the entry words on the page begin with. They appear along the outside margin.

Did You Know? relates interesting facts or ideas about the entry word.

Definitions for each entry word are user friendly for students, parents, and teachers.

Related terms are included where appropriate.

Diagrams are used to illustrate mathematics concepts.

Related Words are mathematics terms that share the same root.

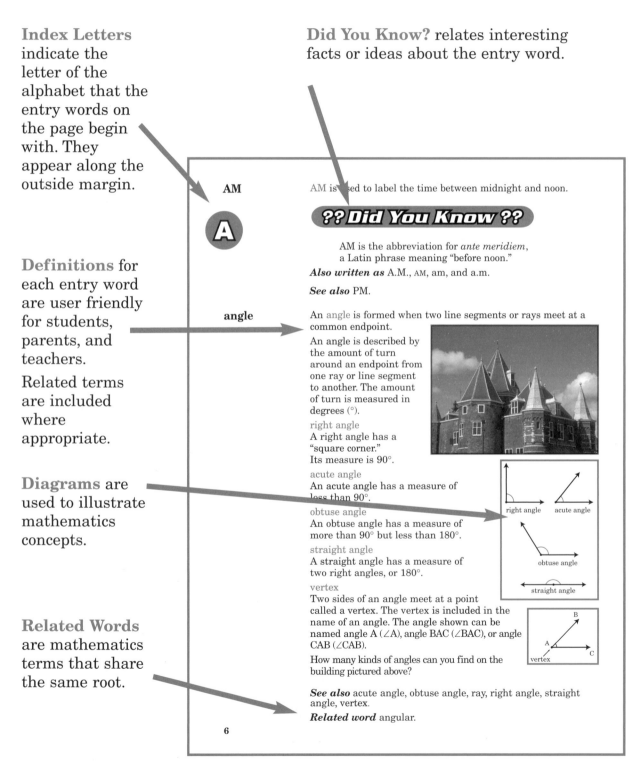

AM

A

AM is used to label the time between midnight and noon.

?? Did You Know ??

AM is the abbreviation for *ante meridiem*, a Latin phrase meaning "before noon."

Also written as A.M., AM, am, and a.m.

See also PM.

angle

An angle is formed when two line segments or rays meet at a common endpoint.

An angle is described by the amount of turn around an endpoint from one ray or line segment to another. The amount of turn is measured in degrees (°).

right angle
A right angle has a "square corner." Its measure is 90°.

acute angle
An acute angle has a measure of less than 90°.

obtuse angle
An obtuse angle has a measure of more than 90° but less than 180°.

straight angle
A straight angle has a measure of two right angles, or 180°.

vertex
Two sides of an angle meet at a point called a vertex. The vertex is included in the name of an angle. The angle shown can be named angle A (∠A), angle BAC (∠BAC), or angle CAB (∠CAB).

How many kinds of angles can you find on the building pictured above?

See also acute angle, obtuse angle, ray, right angle, straight angle, vertex.

Related word angular.

6

Entry Words have been carefully selected for their usefulness in learning mathematics. The word (*continued*) indicates that an entry word has been carried over from the previous page.

Tools of the Trade illustrates tools used in doing mathematics. These tools are also used by professionals in jobs such as design and architecture.

angle
(continued)

Tools of the Trade

A protractor is a tool used to measure angles.

apex
The apex of a cone or pyramid is its highest point in relation to its base.

See also cone, pyramid.

apex   apex

cone   pyramid

approximate
Approximate means almost or near.

Example
The ancient Colosseum, a huge amphitheater built in Rome nearly 2,000 years ago, could hold approximately 50,000 spectators.

See also about.
Related words approximately, approximation.

See also links the entry words to synonyms, similar terms, or other terms in the dictionary that may extend understanding.

approximate number
An approximate number is a number that is close but not exactly equal to another number.

Example
Chad saw this sign at the grocery store: 3 candy bars for $1.00. At this rate, what was the cost of each candy bar?
$1.00 ÷ 3 is a little more than $ .33 and a little less than $ .34. Both $ .33 and $ .34 are approximate numbers for the cost of each bar.
In this case, Chad paid $ .34 for one candy bar.

See also about, approximate.

Photographs and Illustrations have been selected to show real-world connections to many mathematics terms.

arc
An arc is a part of a circle or curve between two points.
This bridge has supports in the shape of an arc.

7

abacus

plural **abaci**

An abacus is a device used for counting or doing operations.

Tools of the Trade

The abacus is an ancient device that is still used in some countries, including China and Japan. In the United States it is often used to help show the concept of place value.

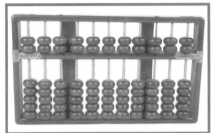

?? Did You Know ??

Shopkeepers in several Asian countries still use the abacus instead of a calculator. It can also be used to help us understand basic arithmetic. Children who have vision difficulties can learn arithmetic using the abacus instead of paper and pencil.

about

About means "almost the same as" or "very close to."

Examples

A stick of gum is about 7 centimeters long.

The sum of $2\frac{1}{2}$ and $\frac{1}{4}$ is about 3.

See also approximate.

acre

An acre is a unit for measuring area in the customary system of measurement.

1 acre = 43,560 ft^2

640 acres = 1 mi^2

?? Did You Know ??

acre
(continued)

An acre was originally a field that a farmer, using a pair of oxen, could plow in a day. By nightfall, the farmer and the team of oxen had walked about 10 miles turning an acre of soil.

See also customary system of measurement.

acute angle

An acute angle is an angle that measures less than 90°.

See also angle.

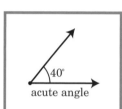

acute triangle

An acute triangle is a triangle with three acute angles, which are angles measuring less than 90°.

See also triangle.

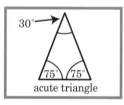

addend

An addend is a number that is to be added to another number.

Examples

In $3 + 6 = 9$,
3 and 6 are addends.

In $1.34 + 7\frac{1}{4} = 8.59$,
1.34 and $7\frac{1}{4}$ are addends.

See also addition.

addition

Addition is a way of finding the amount of two sets or quantities that are put together. It is an arithmetic operation. (The four basic arithmetic operations are addition, subtraction, multiplication, and division.)

See also addition fact, addition sentence.
Related words add, addend, additive, additional.

addition facts

The basic addition facts are the 100 addition combinations of one-digit numbers.

addition facts
(continued)

Examples

$5 + 0 = 5$ $8 + 6 = 14$

See also basic facts.

addition sentence

An addition sentence is a number sentence used to express addition.

Example

$10.3 + 7.2 = 17.5$
In $10.3 + 7.2 = 17.5$,
10.3 and 7.2 are addends.
+ is the symbol for addition.
17.5 is the sum.

See also addend, addition, number sentence, sum.

addition table

An addition table is a way of showing the basic addition facts. It includes sums for the 100 combinations of one-digit numbers.

See also basic facts.

additive inverse

The additive inverse of a number is a number that is the same distance from 0 on the number line, but in the opposite direction.

Example

$^{+}3 + {}^{-}3 = 0$

In the example, $^{+}3$ and $^{-}3$ are additive inverses, or opposites.
$^{-}3$ is the additive inverse, or opposite, of $^{+}3$.
$^{+}3$ is the additive inverse, or opposite, of $^{-}3$.

Also called opposite of a number.
See also additive inverse property, opposite of a number.

additive inverse property

The additive inverse property means that the sum of a number and its opposite is 0.

Also called inverse property of addition.

adjacent

Adjacent is used to describe angles. It is also used to describe the sides of polygons.

Adjacent angles are two angles in a plane that have a common vertex and a common side. They do not have any common interior points. In other words, they do not share any "inside space."

Adjacent sides are two sides of a polygon that have a common vertex.

See also angle, polygon.

algebra

Algebra is a branch of mathematics in which arithmetic is extended to deal with unknown numbers or relationships, using letters or other symbols. These letters or symbols are called variables.

Example

Tyler counted the empty spaces in his stamp book.
He found that he needs to collect only 37 more stamps
to fill his book, which holds 500 stamps.
How many stamps does he have already?

To solve this problem using algebra, first write an equation that describes the problem situation. Let s represent the number of stamps that Tyler already has.

$$s + 37 = 500$$

One way to solve for s is to subtract 37 from each side of the equation.

$$s + 37 - 37 = 500 - 37$$
$$s = 463$$

Tyler already has 463 stamps.

See also variable.
Related word algebraic.

algorithm

An algorithm is a systematic step-by-step procedure.

partial products algorithm
The following procedure is the partial products algorithm for

algorithm
(continued)

multiplication. It is often used for paper-and-pencil multiplication of numbers with two or more digits.

Example

```
    24
  × 13      Step 1: Think of 13 as 10 + 3
    72      Step 2: 3 × 24 = 72
   240      Step 3: 10 × 24 = 240
   312      Step 4: 72 + 240 = 312
```

Related word algorithmic.

altitude

An altitude of a geometric figure is a line segment that shows the figure's height. Altitude is also the length of that line segment. Altitude can also be used to mean elevation, or distance above or below sea level.

Examples

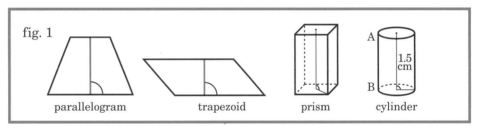

parallelogram   trapezoid   prism   cylinder

The altitudes for these geometric figures are perpendicular to both bases. The altitude of the cylinder (fig. 1) is line segment AB, or 1.5 centimeters.

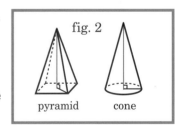

For a pyramid or cone (fig. 2), the altitude is a line segment from the base to the vertex perpendicular to the base.

A triangle has three altitudes because any one of its sides can be a base. (In other words, a triangle can "sit" on any one of its

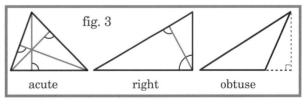

acute   right   obtuse

sides.) All three altitudes are pictured for the acute triangle. Can you find the remaining two altitudes for the right and obtuse triangles?

Also called height.

See also height.

AM

AM is used to label the time between midnight and noon.

angle

An angle is formed when two line segments or rays meet at a common endpoint.

An angle is described by the amount of turn around an endpoint from one ray or line segment to another. The amount of turn is measured in degrees (°).

right angle
A right angle has a "square corner." Its measure is 90°.

acute angle
An acute angle has a measure of less than 90°.

obtuse angle
An obtuse angle has a measure of more than 90° but less than 180°.

straight angle
A straight angle has a measure of two right angles, or 180°.

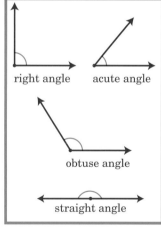

right angle   acute angle

obtuse angle

straight angle

vertex
Two sides of an angle meet at a point called a vertex. The vertex is included in the name of an angle. The angle shown can be named angle A (∠A), angle BAC (∠BAC), or angle CAB (∠CAB).

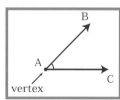

B

A

C

vertex

How many kinds of angles can you find on the building pictured above?

See also acute angle, obtuse angle, ray, right angle, straight angle, vertex.

Related word angular.

angle
(continued)

A protractor is a tool used to measure angles.

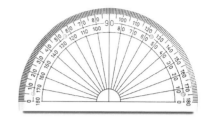

apex

The apex of a cone or pyramid is its highest point in relation to its base.

See also cone, pyramid.

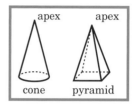

approximate

Approximate means almost or near.

Example

The ancient Colosseum, a huge amphitheater built in Rome nearly 2,000 years ago, could hold approximately 50,000 spectators.

See also about.
Related words approximately, approximation.

approximate number

An approximate number is a number that is close but not exactly equal to another number.

Example

Chad saw this sign at the grocery store: 3 candy bars for $1.00. At this rate, what was the cost of each candy bar?

$1.00 ÷ 3 is a little more than $ .33 and a little less than $ .34. Both $ .33 and $ .34 are approximate numbers for the cost of each bar.
In this case, Chad paid $ .34 for one candy bar.

See also about, approximate.

arc

An arc is a part of a circle or curve between two points.

This bridge has supports in the shape of an arc.

arc
(continued)

Semicircle is the name for an arc that is a half-circle. (One meaning of semi- is "half.")

See also semicircle.

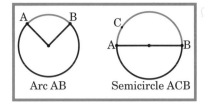

area

The area of a plane figure is the space enclosed by that figure. Area is also the measure of the enclosed space. Area is expressed in square units.

To find the area of plane figures:

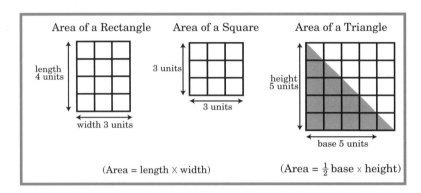

See also cm², ft², in.², km², m², mi², square unit, yd².

arithmetic

Arithmetic is the branch of mathematics that deals with the use of one or more of the basic operations on numbers. These basic operations are addition, subtraction, multiplication, and division.

See also addition, basic operations, division, multiplication, operation, subtraction.

arithmetic mean

The arithmetic mean of a set of numbers is an average. To find the arithmetic mean, first find the sum of the set of numbers, then divide by the number of numbers in the set.

See also average.

arithmetic sequence

An arithmetic sequence is a number pattern in which the difference between any two consecutive numbers is the same.

Example

3, 7, 11, 15, 19, … is an arithmetic sequence.
The difference between any two consecutive numbers is 4.

arithmetic sequence
(continued)

Also called arithmetic progression.

See also pattern.

array

An array can be an orderly arrangement of objects, such as cans on a shelf or seats in a theater. Pictured here is a 4 × 2 array of sushi.

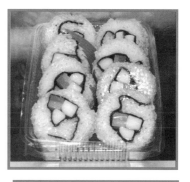

An array can also be a drawing. Arrays drawn in the shape of a rectangle are often used to model multiplication.

Pictured here is a 2 × 4 array.

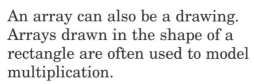

associative property

The associative property means that changing the grouping of the numbers used in an operation does not change the result of that operation. Addition and multiplication have the associative property, but subtraction and division do not.

Also called grouping property.

See also associative property of addition, associative property of multiplication.

associative property of addition

The associative property of addition means that when adding three or more numbers, the way the numbers are grouped will not change the result. The sum will remain the same.

Example

6 + 3 + 7 =	6 + 3 + 7 =
(6 + 3) + 7 =	6 + (3 + 7) =
9 + 7 = 16	6 + 10 = 16

The associative property of addition is often written with symbols as

$$(a + b) + c = a + (b + c)$$

Also called grouping property of addition.

See also associative property of multiplication.

associative property of multiplication

The associative property of multiplication means that when multiplying three or more numbers, the way the numbers are grouped will not change the result. The product will remain the same.

Example

$4 \times 2 \times 5 =$	$4 \times 2 \times 5 =$
$(4 \times 2) \times 5 =$	$4 \times (2 \times 5) =$
$8 \times 5 = 40$	$4 \times 10 = 40$

The associative property of multiplication is written with symbols as

$$(a \times b) \times c = a \times (b \times c)$$

Also called grouping property of multiplication.

See also associative property of addition.

attribute

The attributes of an object are its characteristics or qualities. Some attributes can best be described with words, some by counting, and others by measuring.

Example

The mother bear has newborn cubs (words).

There are three of them (counting).

Each one weighs about $\frac{1}{2}$ pound at birth (measuring).

Also called property.

average

The average is a single number used to represent a set of numbers. We use three kinds of averages: the arithmetic mean (usually called the mean), the median, and the mode.

mean

The mean is the most familiar kind of average.

Example

Cheryl earned the following points on math tests:
  (88, 88, 43, 96, 85).

To find the mean, find the total points scored:
  (88 + 88 + 43 + 96 + 85 = 400).

Then divide the total points by the number of tests taken.
  ($400 \div 5 = 80$)

The mean of Cheryl's test scores is 80.

Also called arithmetic mean.

average
(continued)

median

The median is the middle number for a set of data when the data are ordered from least to greatest or greatest to least.

Example

To find the median of Cheryl's test scores, arrange the points in numerical order (43, 85, 88, 88, 96). The median is the middle number, 88.

mode

The mode for a set of numbers is the number or numbers that occur most often.

Example

The mode for Cheryl's test scores is 88 because it occurs twice, and each of the other scores occurs only once.

?? Did You Know ??

The average length lifespan in the United States is nearly 80 years. The average American watches television about 4 hours a day—think about how long that would be in a lifetime!

See also mean, median, mode.

Related words averaged, averaging.

axis

plural **axes**

An axis is a reference line in a coordinate system. The *x*- and *y*-axes are marked in this coordinate system.

See also coordinate system, *x*-axis, *y*-axis.

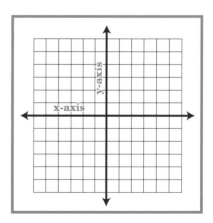

Bb

bar graph

A bar graph (fig. 1) is a kind of graph that we use to compare categories or groups of information. Bar graphs are usually formed with rectangular bars, arranged either vertically or horizonally, to show information. They can also be formed with real objects, pictures, or symbols.

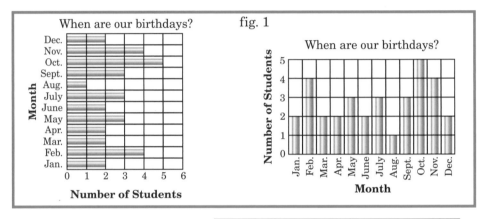

fig. 1

double bar graph

A double bar graph (fig. 2) presents two sets of data on the same graph.

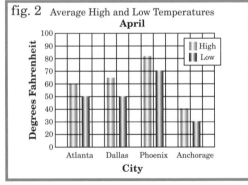

real graph

A real graph (fig. 3) is a kind of bar graph that displays the real objects being graphed.

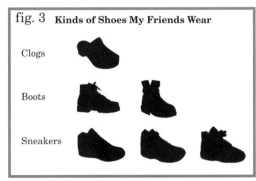

bar graph
(continued)

picture graph

A picture graph is a kind of bar graph that displays pictures or drawings to represent data. "Our Favorite Sports" is an example of a picture graph.

Also called pictograph.

Our Favorite Sports	
Baseball	☺☺☺☺
Football	☺☺
Soccer	☺☺☺
Bowling	☺☺
Dance	☺☺☺☺
Other	☺

☺ = 2 students

symbolic graph

A symbolic graph uses a symbol such as an X or an O or the blocks on the graph to show information. "Seedling's Growth" is an example of a symbolic graph.

See also line plot, scale.

Seedling's Growth	
June	✔
July	✔✔
August	✔✔✔
September	✔✔✔✔✔
October	✔✔✔✔✔✔✔

✔ = 2

base

base of a plane figure

A base of a plane figure is usually thought of as a side on which the figure "sits."

Examples

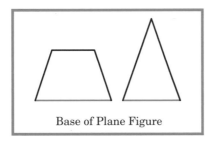

Base of Plane Figure

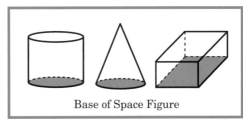

Base of Space Figure

base of a solid figure

A base of a solid figure is usually thought of as a face upon which it can "sit." One base is shown for each of the plane and space figures, but most figures have more than one base.

base of a numeration system

The base of a numeration system is the number that is used to build the place value of that system. Our decimal numeration system is a base-ten numeration system.

base
(continued)

base of an exponent

The base of an exponent is a number that is raised to a power.

In 10^2, 10 is the base and 2 is the exponent.
10 is raised to the power of 2.
($10^2 = 10 \times 10 = 100$)

Base-ten blocks are place-value materials. They are tools for helping us see relationships among the places in the decimal, or base-ten, numeration system.

See also decimal numeration system, exponent, power of a number.

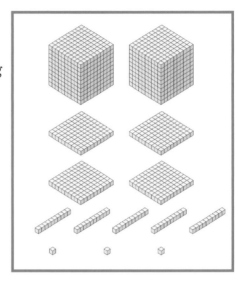

base-ten numeration system

The base-ten numeration system is a system of grouping by tens. For example, when 9 ones have been counted, the next number will be recorded as one group of tens and zero ones (10).

Also called decimal numeration system.

basic facts

The basic facts are all the addition and multiplication combinations of one-digit numbers. There are 100 basic addition facts. There are also 100 basic multiplication facts.

Also called basic number facts.

See also commutative property of addition, commutative property of multiplication, fact family, related facts.

+	0	1	2	3	4	5	6	7	8	9
0	0	1	2	3	4	5	6	7	8	9
1	1	2	3	4	5	6	7	8	9	10
2	2	3	4	5	6	7	8	9	10	11
3	3	4	5	6	7	8	9	10	11	12
4	4	5	6	7	8	⑨	10	11	12	13
5	5	6	7	8	⑨	10	11	12	13	14
6	6	7	8	9	10	11	12	13	14	15
7	7	8	9	10	11	12	13	14	15	16
8	8	9	10	11	12	13	14	15	16	17
9	9	10	11	12	13	14	15	16	17	18

×	0	1	2	3	4	5	6	7	8	9
0	0	0	0	0	0	0	0	0	0	0
1	0	1	2	3	4	5	6	7	8	9
2	0	2	4	6	8	10	12	14	16	18
3	0	3	6	9	12	15	18	㉑	24	27
4	0	4	8	12	16	20	24	28	32	36
5	0	5	10	15	20	25	30	35	40	45
6	0	6	12	18	24	30	36	42	48	54
7	0	7	14	㉑	28	35	42	49	56	63
8	0	8	16	24	32	40	48	56	64	72
9	0	9	18	27	36	45	54	63	72	81

$4 + 5 = 9$ $5 + 4 = 9$ $3 \times 7 = 21$ $7 \times 3 = 21$

basic operations

The four basic operations of arithmetic are addition, subtraction, multiplication, and division.

See also addition, arithmetic, basic facts, division, multiplication, operation, subtraction.

benchmark

A benchmark is a reference point that can be used to help make an estimate.

Example

In adding 8+9, 10 and 20 are good benchmark numbers for making an estimate.

See also estimate.

billion

A billion is equal to 1,000 millions.
In standard form, one billion is written as 1,000,000,000.

With an exponent, one billion may be written as 1×10^9 (or simply as 10^9).

See also decimal numeration system, symbols.

bisect

To bisect is to divide a geometric figure such as a line or angle into two parts that are the same size and shape.

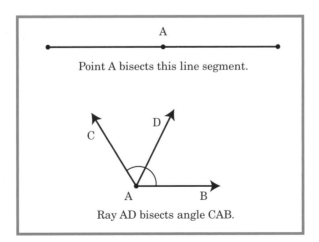

Point A bisects this line segment.

Ray AD bisects angle CAB.

Related words bisector, bisecting lines.

capacity

Capacity usually refers to the amount of liquid a container can hold. It is reported in units of liquid measure.

Capacity is also used in other measurements. For example, an elevator may have a capacity of 13 people or 2,100 pounds.

See also cup, gallon, kiloliter, liter, milliliter, pint, quart, tablespoon, teaspoon.

cardinal number

A cardinal number is a whole number that represents how many are in a group.

Example

A soccer team has 11 players on the field at one time. (11 is a cardinal number because it tells how many players there are.)

Cartesian coordinate system

The Cartesian coordinate system is a system used to locate points in a plane relative to the intersection of two perpendicular lines, or *axes*, in the plane.

?? Did You Know ??

The story is told that Descartes, a French philosopher and mathematician of the 17th century, invented the rectangular coordinate system (Cartesian coordinate system) as he watched a fly crawl across the ceiling while he was sick in bed. Later, the Cartesian coordinate system was expanded to include three dimensions. (How are the words *Descartes* and *Cartesian* alike?)

Also called rectangular coordinate system.
See also coordinate system, x- and y-axes.

Celsius (°C) temperature scale

The Celsius (°C) temperature scale is used for measuring temperature in the metric system.

Two reference points for this temperature scale are
      0° freezing point of water
      100° boiling point of water

Also called Celsius scale, centigrade scale.

See also temperature.

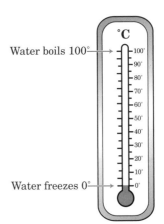

Water boils 100°
Water freezes 0°
°C

center

center of a circle

The center of a circle is a point inside a circle that is an equal distance from all points on the circle.

center

center of a sphere

The center of a sphere is a point inside a sphere that is an equal distance from all points on the sphere.

See also circle, sphere.

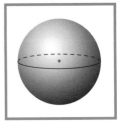

centi-

Centi- is a prefix meaning one hundredth.
      1 centimeter = 0.01 meter
      1 centiliter = 0.01 liter
      1 centigram = 0.01 gram

centigrade scale

The centigrade scale is a name sometimes used for the Celsius temperature scale. The name *centigrade* comes from the 100 degrees, or units, between the freezing and boiling points of water.

Also called Celsius (°C) temperature scale.

See also Celsius (°C) temperature scale.

centigram (cg)

A centigram is a unit of weight in the metric system of measurement.
      100 centigrams = 1 gram

See also metric system of measurement.

centiliter (cL)

A centiliter is a unit of capacity in the metric system of measurement.

> 100 centiliters = 1 liter

See also metric system of measurement.

centimeter (cm)

A centimeter is a unit of length in the metric system of measurement.

> 100 centimeters = 1 meter

Example

> The width of a child's pinkie finger is a good benchmark for a centimeter.

See also metric system of measurement.

century

A century is a measure of time.

> 1 century = 100 years

?? Did You Know ??

The United States has been a country for a little more than two centuries. The oldest country in the world is China, which has a recorded history of about 30 centuries.

chance

Chance is the likelihood that a given event will occur.

Also called probability.
See also probability.

chart

A chart is a form used to record information.

chord

A chord is any line segment with both endpoints on a circle.

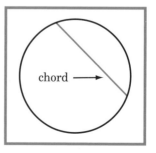

18

chord
(continued)

diameter
The diameter is a special kind of chord that passes through the center of a circle.

See also circumference, diameter, radius.

circle

A circle is a closed plane figure with all points the same distance from the center.

?? Did You Know ??

The region between two concentric circles is called an annulus. The name comes from a Latin root meaning "little ring." *Annulus* is also used to name ring-like structures such as the yearly growth ring of a tree.

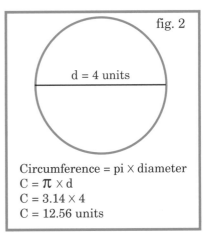

circle graph

A circle graph (fig.1) is a graph in the shape of a circle, or pie. It shows how the total amount has been divided.

Also called pie chart, pie graph.
See also bar graph, line graph.

Favorite Fast Foods

Chinese food

Hamburgers

Pizza

Mexican food

fig. 1

circumference

Circumference is the distance around a circle, which equals a little more than three times its diameter. (fig. 2)

pi (π)
The actual ratio of the circumference of a circle to its diameter is known as pi, written as π. The approximate value of π is 3.14. To find the circumference (C) of a circle, multiply the diameter (d) of the circle by 3.14 (an approximation of π). The formula for finding the circumference of a circle may be written as: $C = \pi \times d$ or $C = \pi d$

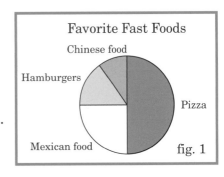

fig. 2

d = 4 units

Circumference = pi $\times$ diameter
$C = \pi \times d$
$C = 3.14 \times 4$
$C = 12.56$ units

See also perimeter, pi.

clockwise

Clockwise is the direction in which the hands of a clock move. The numbers are passed in order from least to greatest.

See also counterclockwise.

closed curve

A closed curve is a curve that flows continuously with no breaks or gaps.

simple closed curve
A simple closed curve is a closed curve that does not cross itself. The three shapes on the left are simple closed curves.

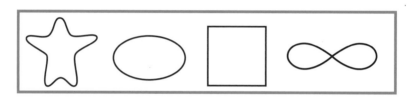

Also called closed figure.

clustering

Clustering is a method used for estimating a result when numbers appear to group, or cluster, around a common number.

Example

Juan bought decorations for a party. He spent $3.63 for balloons, $3.85 for party favors, and $4.55 for streamers. About how much did he spend for decorations?

$3.63, $3.85, and $4.55 cluster around $4.
4 + 4 + 4 = 12 (or 3 × 4 = 12)

Juan spent about $12 for party decorations.

See also estimation strategies.

cm²

Read as *square centimeter*.

A cm² is equal to the area of a square that measures 1 centimeter on each side.

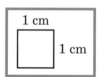

Also written as square centimeter.

cm²
(continued)

See also square, square unit.

cm³

Read as *cubic centimeter*.

A cm³ is equal to the volume of a cube that measures 1 cm in length, 1 cm in width, and 1 cm in height.

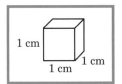

Also written as cubic centimeter.
See also cube, cubic unit.

column

A column is a vertical (from top to bottom) arrangement of items or numbers in an array or table.

See also array, table.

Year	Number of Missions With Astronauts	Time in Space
1961	🚶🚶	30 minutes
1962	🚶🚶🚶	19 hours
1963	🚶	34 hours
1964		0
1965	🚶🚶🚶🚶	330 hours
1966	🚶🚶🚶🚶	309 hours
1967		0
1968	🚶🚶🚶	407 hours
1969	🚶🚶	873 hours

Manned U.S. Space Travel in 1960s

↑ column   🚶 = one mission with astronauts

common denominator

A common denominator is a common multiple of the denominators of two or more fractions. Two of the common denominators for $\frac{1}{6}$ and $\frac{3}{8}$ are 24 and 48. The least common denominator is 24.

See also least common denominator.

common factor

A common factor is a factor that two or more numbers share.

Example

Factors of 10: 1, 2, 5 and 10
Factors of 20: 1, 2, 4, 5, 10, and 20
Common factors of 10 and 20 include 1, 2, 5, and 10.
The greatest common factor of 10 and 20 is 10.

Also called common divisor.
See also greatest common factor.

common fraction

A common fraction is one way of expressing a fractional number. (Decimal fractions and percents are two other ways of expressing fractional numbers.)

part of a whole or set
A common fraction may name a part of a whole or a set.

Examples

$\frac{3}{4}$ of the pie is left. $\frac{3}{4}$ is a common fraction naming a part of the whole pie.

division
Common fractions are also used to express division.

Example

Nancy wants to make omelets for breakfast. She has 6 eggs. She needs 2 eggs for each omelet. How many omelets can she make?

$$\frac{6}{2} = 3$$

Nancy can make 3 omelets.

ratio
A common fraction can be used to name a ratio.

Example

Dante's birthday party included party hats and noisemakers for each guest. For each guest, there is one hat and two noisemakers. The ratio of party hats to noisemakers can be expressed as the common fraction $\frac{1}{2}$ (or as the ratio 1:2).

In the example at right, the ratio of bees to hive can be expressed as the common fraction $\frac{4}{1}$ (or as the ratio 4:1).

Also called fraction.

See also fraction, fractional number, rational number.

common multiple

A common multiple is a multiple that two or more numbers share.

Example
Multiples of 4: 4, 8, 12, 16, 20, 24 …
Multiples of 8: 8, 16, 24, …
Some of the common multiples of 4 and 8 are 8, 16, and 24.

least common multiple (LCM)
The least common multiple is the smallest multiple other than 0 that two or more numbers share.

The least common multiple of 4 and 8 is 8.
This is written as LCM (4, 8) = 8.

See also least common multiple, multiple.

commutative property

The commutative property means that changing the order of the two numbers used in an operation does not change the result of that operation. Addition and multiplication have the commutative property, but subtraction and division do not.

Also called order property.
See also commutative property of addition, commutative property of multiplication.

commutative property of addition

The commutative property of addition means that changing the order in which two numbers are added does not change the sum.

Examples
$$7 + 9 = 16 \qquad \frac{1}{4} + \frac{1}{2} = \frac{3}{4}$$
$$9 + 7 = 16 \qquad \frac{1}{2} + \frac{1}{4} = \frac{3}{4}$$
The commutative property of addition is often written with symbols as
$$a + b = b + a$$

Also called order property of addition.
See also commutative property, commutative property of multiplication.

commutative property of multiplication

The commutative property of multiplication means that changing the order in which two numbers are multiplied does not change the product.

Examples

8 × 5 = 40	1.3 × 4 = 5.2
5 × 8 = 40	4 × 1.3 = 5.2

The commutative property of multiplication is often written with symbols as

$$a \times b = b \times a.$$

Also called order property of multiplication.

See also commutative property of addition.

compass

A compass (fig. 1) is a tool that can be used to draw circles and arcs and to copy line segments. Many plane figures can be constructed with only a compass and a straightedge.

Another kind of compass (fig. 2) is used as a tool for finding direction.

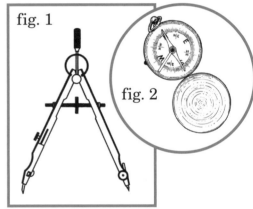

fig. 1

fig. 2

See also straightedge.

compatible numbers

Compatible numbers are numbers that seem to "go together" because they are easy to compute mentally. The use of compatible numbers is an estimation strategy.

Example

Bob bought a package of 4 pairs of athletic socks for $10.99. About how much did he pay for each pair?

$10.99 is close to $12.00 (12 is a multiple of 4).

12 ÷ 4 = 3

Each pair of socks cost about $3.00.

See also estimation strategies.

complete factorization

Complete factorization is the expression of a composite number as the product of its prime factors.

Example $24 = 2 \times 2 \times 2 \times 3 = 2^3 \times 3$

Also called prime factorization.

See also composite number, prime factor, prime factorization.

composite number

A composite number is a whole number with more than two factors.

Example

16 is a composite number. Its factors are 1, 2, 4, 8, and 16. These arrays for 16 show its factors.

```
1 □□□□□□□□□□□□□□□□
           16

2 □□□□□□□□        4 □□□□
  □□□□□□□□          □□□□
      8             □□□□
                    □□□□
                     4
```

compound event

A compound event is two or more independent events considered together.

independent events
Independent events are events that have no effect on each other.

Example

You are playing a game in which you are rolling two dice. You score if the sum of the numbers rolled on your dice is an even number. Rolling one die and rolling the other die are two independent events. The results of both dice are considered together to determine if you rolled an even number. This is a compound event.

See also dependent event, independent events.

computation

Computation is doing the arithmetic operations needed to solve a problem.

See also algorithm, basic operations, mental computation.
Related word computer.

concave

A common meaning for concave is "curved inward" like the inside of a bowl or the inner surface of a sphere.

concave⟶

concave polygon
A concave polygon is a polygon with one or more of the vertices "arched inward." A polygon is concave if there is at least one pair of points within the figure that could be joined by a line segment that would go outside the figure.

Also called nonconvex.
See also convex.

concentric circles

Concentric circles have different diameters but have the same center. A pebble dropped in a pond creates concentric circles, as does this milk drop pictured at right.

cone

A cone is a space figure that usually has a circular base. It comes to a point opposite the base.

apex
The point, or vertex, is called the apex.

apex

base

Related word conical.

congruent figures

Congruent figures are figures that have the same size and shape.

congruent figures
(continued)

Examples

Which pairs of figures pictured here are congruent because they are the same size and shape?

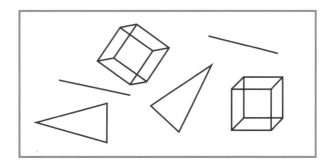

Identical twins have congruent, or almost congruent, features.

construction

A construction is a drawing of a geometric figure made by using only a compass and a straightedge.

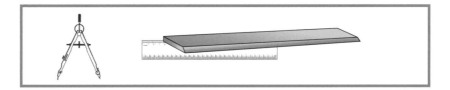

A compass is used for drawing circles and arcs and for copying line segments. The straightedge is used for drawing line segments.

See also compass, straightedge.

convex

A common meaning for convex is curved outward. The outside of a bowl and the outer surface of a sphere are convex.

←convex

convex polygon
A convex polygon is a polygon with all of its vertices "arched outward." A polygon is convex when each pair of points within the figure could be joined by a line segment that would not go outside the figure.

See also concave.

coordinates

Coordinates in a plane are the two items in an ordered pair, used to identify a location on a map or in a coordinate plane. These coordinates are relative to a fixed point or origin.

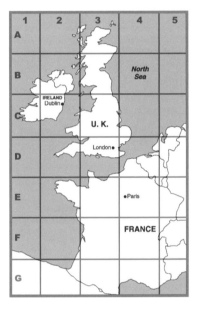

Examples

Often maps will use a letter as one of the coordinates for locating a point and a numeral for the other.

Coordinates on this map include:

Dublin, Ireland	C 2
London, England	D 3
Paris, France	E 4

The coordinates of the point on this plane are 6 and 4. The first coordinate, 6, is the x-coordinate. The second coordinate, 4, is the y-coordinate.

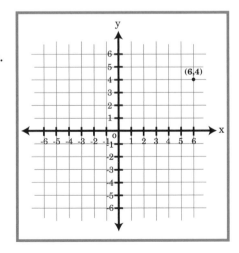

See also coordinate system, origin.

coordinate system

A coordinate system is used for locating points. When used to locate points on a plane, it is also called a rectangular coordinate system.

x-axis
The *x*-axis is the horizontal number line.

y-axis
The *y*-axis is the vertical number line.

origin
The origin is the point of intersection of the *x*- and *y*-axes.

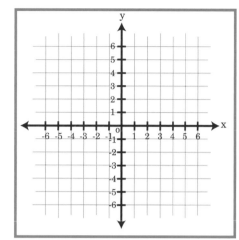

?? Did You Know ??

There are many other coordinate systems used today. One of the most exciting is the global positioning system (GPS), which allows you to find your location anywhere on Earth at any time, regardless of the weather.

quadrant
The *x*- and *y*-axes divide the rectangular coordinate system into four sections, or quadrants. They are numbered in order from I to IV, starting in the upper right quadrant and going counterclockwise.

An ordered pair, for example, names a point and tells its location in the coordinate system.

x-coordinate
The first number in the ordered pair is called the *x*-coordinate. It indicates a distance along the *x*-axis.

y-coordinate
The second number in the ordered pair is called the *y*-coordinate. It indicates a distance along the *y*-axis.

plotting a point
Plotting a point, or locating and marking a point when given its coordinates, is described in these examples.

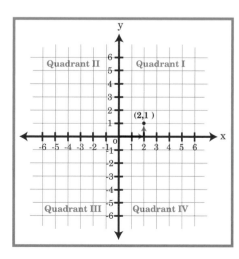

coordinate system
(continued)

Examples

To plot (4,3), first locate 4 on the *x*-axis and then 3 on the *y*-axis. (Follow the arrows shown in the drawing.) Because both coordinates are positive numbers, the point is located in Quadrant I.

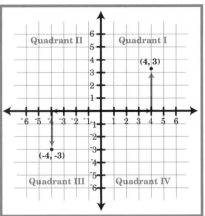

To plot (⁻4,⁻3), first locate ⁻4 on the *x*-axis and ⁻3 on the *y*-axis. (Follow the arrows shown in the drawing.) Because both coordinates are negative numbers, the point is located in Quadrant III.

(4,⁻3) locates a point in Quadrant IV.

(⁻4, 3) locates a point in Quadrant II.

counter-clockwise

Counterclockwise is the direction that is the opposite of the way the hands of a clock move normally.

counting numbers

Counting numbers are the numbers 1, 2, 3, 4, 5, and so on. They go on without end. (0 and negative numbers are not included in counting numbers.) Counting numbers are used to name the number in a set of objects.

Also called natural numbers.

counting principle

The counting principle is a way to find the possible choices or outcomes in situations for which order is important.

Examples

There are 3 kinds of sandwiches and 2 flavors of ice cream in the refrigerator. You may choose one kind of sandwich and then one flavor of ice cream for lunch. How many different choices do you have for lunch?

> 3 (sandwiches) × 2 (flavors of ice cream) = 6
> There are 6 choices.

counting principle
(continued)

A coin is tossed 3 times. Each outcome is either heads or tails. How many different outcomes are possible?

2 = number of possible outcomes for the first toss
2 = number of possible outcomes for the second toss
2 = number of possible outcomes for the third toss
$2 \times 2 \times 2 = 8$

8 different outcomes are possible.

Also called basic counting principle, fundamental counting principle.

See also tree diagram.

cube

A cube is a space figure with 6 square faces. These faces are all the same size.

Also called hexahedron.

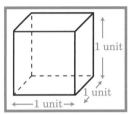

It is one of the five regular polyhedra.

The cube of a number is that number used as a factor 3 times.

Examples

$2^3 = 2 \times 2 \times 2 = 8$
$10^3 = 10 \times 10 \times 10 = 1,000$

See also polyhedron.
Related words cubic, cuboid.

cubic centimeter (cm³)

A cubic centimeter is equal to the volume of a cube that measures 1 centimeter on each edge.

See also cubic unit, cm³.

cubic decimeter (dm³)

A cubic decimeter is equal to the volume of a cube that measures 1 decimeter (10 cm) on each edge.

See also cubic unit, dm³.

cubic foot (ft³)

A cubic foot is equal to the volume of a cube that measures 1 foot on each edge.

See also cubic unit, ft³.

cubic inch (in.³)

A cubic inch is equal to the volume of a cube that measures 1 inch on each edge.

Example

An ice cube is about the size of 1 in.³

See also cubic unit, in.³.

cubic meter (m³)

A cubic meter is equal to the volume of a cube that measures 1 meter on each edge.

See also cubic unit, m³.

cubic unit

A cubic unit is a unit for measuring volume. Each face of a cubic unit is a square, and each edge is one unit in length.

In the metric system, commonly used cubic units include cm³ (cubic centimeter), dm³ (cubic decimeter), and m³ (cubic meter).

In the customary system, commonly used cubic units include in.³ (cubic inch), ft³ (cubic foot), and yd³ (cubic yard).

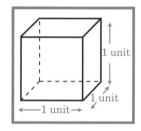

cubic yard (yd³)

A cubic yard is equal to the volume of a cube that measures 1 yard on each edge.

?? Did You Know ??

Although a cubic unit of volume is usually thought of in the shape of a cube, it can also take other shapes. For example, a cubic yard of concrete may be poured and spread a few inches deep to form a small patio.

See also customary system of measurement, metric system of measurement.

cuboid

A cuboid is a prism that has all rectangular faces. Its angles are all right angles and opposite faces are equal.

If all the faces are square, it is usually known as a cube. If the bases of a cuboid are square, it is also called a square prism.

Also called right rectangular prism, rectangular parallelopiped.

cup (C)

A cup is a unit of capacity in the customary system of measurement. It holds 8 fluid ounces.

> 1 cup = 8 fluid ounces
> 2 cups = 1 pint

See also capacity, customary system of measurement, fluid ounces.

customary system of measurement

The customary system of measurement is the measurement system used most commonly in the United States. The customary system was brought to this land by colonists from England. Many cultures, some of them dating back to ancient times, contributed to its development. The customary system includes units for measuring length, weight, capacity, area, volume, and temperature.

length
Commonly used units for measuring length in the customary system include inch (in.), foot (ft), yard (yd), and mile (mi).

> 12 in. = 1 ft
> 3 ft = 1 yd
> 5,280 ft or 1,760 yd = 1 mi

weight
Commonly used units for measuring weight in the customary system include ounce (oz), pound (lb), and ton (T).

> 16 oz = 1 lb
> 2,000 lb = 1 T

**customary
system of
measurement**
(continued)

capacity

Commonly used units for measuring capacity in the customary system include teaspoon (t), tablespoon (T), cup (C), pint (pt), quart (qt), and gallon (gal).

3 t = 1 T	2 pt = 1 qt
16 T = 1 C	4 qt = 1 gal
2 C = 1 pt	

fluid ounces (fl oz) are also used for measuring capacity.

1 fl oz = 2 T	16 fl oz = 1 pt
8 fl oz = 1 C	32 fl oz = 1 qt

area

Square units are used for measuring area. In the customary system these include in.^{2}, ft^2, and yd^2.

$$144 \text{ in.}^2 = 1 \text{ ft}^2 \qquad 9 \text{ ft}^2 = 1 \text{ yd}^2$$

Acres and mi^2 are also used for measuring area.

$$1 \text{ acre} = 43{,}560 \text{ ft}^2 \qquad 640 \text{ acres} = 1 \text{ mi}^2$$

volume

Cubic units are used for measuring volume. In the customary system these include in.^{3}, ft^3, and yd^3.

$$1728 \text{ in.}^3 = 1 \text{ ft}^3 \qquad 27 \text{ ft}^3 = 1 \text{ yd}^3$$

temperature

The Fahrenheit (°F) scale is used for measuring temperature in the customary system. Two reference points for this temperature scale are

32° F	the freezing point of water
212°F	the boiling point of water

Also called customary system, customary measurement system, English measurement system, English system of measurement, standard system of measurement, and U.S. Customary System.

See also area, capacity, cubic unit, cup, Fahrenheit (°F) temperature scale, fluid ounce, foot, ft^2, ft^3, gallon, in.^{2}, in.^{3}, inch, length, metric system of measurement, mi^2, mile, ounce, pint, pound, quart, square unit, tablespoon, teaspoon, temperature, ton, volume, weight, yard, yd^2, yd^3.

cylinder

A cylinder is a space figure with two parallel bases (usually circles) that are the same size. A tracing of the net in the drawing can be cut and taped to form a cylinder.

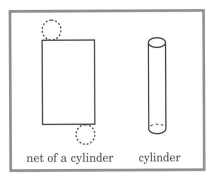

net of a cylinder   cylinder

Examples

These photos illustrate cylinders used for a pipe organ, a tanker truck, and industrial tanks.

Related word cylindrical.

data

Singular **datum**

Data are facts or information gathered for a purpose. Data may be in the form of either words or numbers. This sign gives us information, or data, about the movie being shown. The name of the movie and its rating are in words or letters. The times are shown as numbers.

TWO FRENCH POODLES
G   7:15   9:30

statistics

The branch of mathematics called statistics provides us with tools for organizing, representing, analyzing, and interpreting data. Data are often displayed in tables and graphs.

See also statistics.

Average Gestation Period *(in days)*	
Animal	Number of Days
hamster	16
rabbit	31
squirrel	44
dog	62
lion	108
human	265
horse	337
finback whale	365
giraffe	425
rhinoceros	480
Asian elephant	645

deca-

Deca- is a prefix meaning ten. Also spelled deka-.

See also deka-.

decade

A decade is a measure of time equal to 10 years.

?? Did You Know ??

Decade comes from a Greek word meaning ten. Other words that share the same root include *decathlon* and *decagon*.

decagon   A decagon is a polygon with 10 sides (fig.1). (*deca-* means ten.)

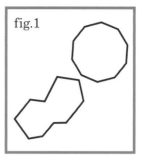

fig.1

deci-   *Deci-* is a prefix meaning one tenth.
        1 decimeter = 0.1 meter
        1 deciliter = 0.1 liter
        1 decigram = 0.1 gram

decigram (dg)   A decigram is a unit of weight in the metric system of measurement.
        10 decigrams = 1 gram

See also metric system of measurement.

deciliter (dL)   A deciliter is a unit of capacity in the metric system of measurement.
        10 deciliters = 1 liter

See also metric system of measurement.

decimal   Decimal means based on ten. The numeration system we use most of the time is called the decimal numeration system because it is based on groupings of ten. Decimal is also used as another name for decimal fraction and decimal mixed number.

See also decimal fraction, decimal mixed number, decimal numeration system.

decimal fraction   A decimal fraction is a fractional number with a denominator of ten or a power of ten. (Common fractions, percents, and decimal fractions are all ways of expressing fractional numbers.)

Examples
Each of the place-value charts in fig. 2 shows a decimal fraction. A decimal fraction is usually written with a decimal point. It can also be expressed as a common fraction with a denominator of ten or a power of ten.

fig.2

ones	.	tenths
0	.	1

0.1 or $\frac{1}{10}$
one tenth

ones	.	tenths	hundreths
0	.	0	1

0.01 or $\frac{1}{100}$
one hundreth

ones	.	tenths	hundreths	thousandths
0	.	0	0	1

0.001 or $\frac{1}{1,000}$
one thousandth

0.8 (or $\frac{8}{10}$)

decimal fraction
(continued)

Examples

$0.1 = \frac{1}{10}$ $0.001 = \frac{1}{1,000}$

$0.01 = \frac{1}{100}$ $0.56 = \frac{56}{100}$

$1.6 = 1\frac{6}{10}$ $0.002 = \frac{2}{1,000}$

Also called decimal.

See also decimal numeration system.

decimal mixed number

A decimal mixed number is made up of a whole number and a decimal fraction.

Example

4.4 inches of rain fell in June, and 1.05 inches fell in July.
4.4 and 1.05 are decimal mixed numbers.

Dollars and cents are expressed in decimal mixed numbers.

Example

Jerry's baseball cap cost $8.95.
8.95 is a decimal mixed number.

Also called decimal.

decimal numeration system

The decimal numeration system is a system for expressing the value of numbers based on grouping by tens.

Each place in the decimal numeration system has a value that is a power of 10. The chart "Whole Numbers" can be continued to the left for writing whole numbers larger than billions. The chart "Decimal Fractions" can be continued to the right for writing decimal fractions smaller than millionths.

Also called base-ten numeration system, Hindu-Arabic numeration system.

See also decimal fraction, place value.

Whole Numbers

Billions			Millions			Thousands			Ones		
hundreds	tens	ones	hundreds	tens	ones	hundreds	tens	ones	hundreds	tens	ones
10^{11}	10^{10}	10^9	10^8	10^7	10^6	10^5	10^4	10^3	10^2	10^1	10^0

Decimal Fractions

ones	.	tenths	hundreths	thousandths	ten-thousandths	hundred-thousandths	millionths
10^0	.	10^{-1}	10^{-2}	10^{-3}	10^{-4}	10^{-5}	10^{-6}

decimal point

A decimal point is the dot used in writing a decimal fraction or a decimal mixed number.

In decimal fractions, 0 is usually placed before the decimal point.

Examples

0.4   Read as *four tenths.*
0.003   Read as *three thousandths.*

In a decimal mixed number, the decimal point separates the whole number from the decimal fraction. The decimal point is read as *and.*

Examples

1.3   Read as *one and three tenths.*
4.58   Read as *four and fifty-eight hundredths.*

A special use of decimal fractions and decimal mixed numbers is in expressing amounts of money. The decimal point is used to separate dollars and cents.

Examples

$0.59   Read as *fifty-nine cents.*
$2.34   Read as *two dollars and thirty-four cents.*

decimeter (dm)

A decimeter is a unit of length in the metric system of measurement.

10 centimeters = 1 decimeter
10 decimeters = 1 meter

See also metric system of measurement.

degree (°)

Degree is the name used for the basic units of measurement for angles and temperature. The angle in this circle is 5°.

angles
The measure of the angle shown is 80°.

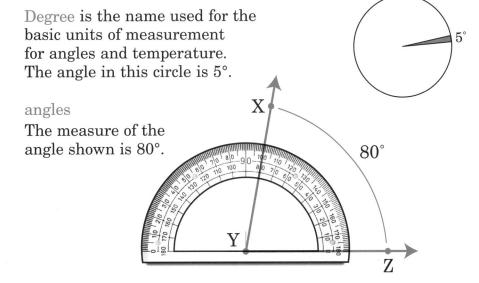

degree (°)
(continued)

temperature

One degree of temperature on the Celsius scale is equal to 1.8 degrees of temperature on the Fahrenheit scale.

degrees Celsius (°C)

The Celsius temperature scale measures temperature in degrees Celsius (°C).
Freezing temperature for water is 0°C.
Boiling temperature for water is 100°C.

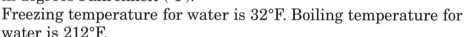

degrees Fahrenheit (°F)

The Fahrenheit temperature scale measures temperature in degrees Fahrenheit (°F).
Freezing temperature for water is 32°F. Boiling temperature for water is 212°F.

degrees latitude and longitude

A degree is also a unit of measure for geographical location.

See also angle, Celsius (°C) temperature scale, Fahrenheit (°F) temperature scale, latitude, longitude.

deka-

Deka- is a prefix meaning ten.
> 1 dekameter = 10 meters
> 1 dekaliter = 10 liters
> 1 dekagram = 10 grams

Also spelled deca-.

dekagram (dag)

A dekagram is a unit of weight in the metric system of measurement.
> 1 dekagram = 10 grams

Also spelled decagram.
See also metric system of measurement.

dekaliter (daL)

A dekaliter is a unit of capacity in the metric system of measurement.
> 1 dekaliter = 10 liters

Also spelled decaliter.
See also metric system of measurement.

dekameter (dam)

A dekameter is a unit of length in the metric system of measurement.

      1 dekameter = 10 meters

Also spelled decameter.
See also metric system of measurement.

?? Did You Know ??

Many classrooms are about 1 dekameter wide.

denominator

The denominator is the name of one of the two terms of a common fraction. It is the part of the fraction that tells how many fractional parts there are in the whole, or set.
It also has other meanings. For example, it is the second term in a ratio.

Notice that the denominator is written below the fraction bar.

$$\dfrac{3}{4} \quad \begin{matrix} \leftarrow \text{numerator} \\ \leftarrow \text{fraction bar} \\ \leftarrow \text{denominator} \end{matrix}$$

Example
In this picture of half an apple, $\frac{1}{2}$ is the fraction represented and 2 is the denominator of the fraction.

See also numerator.

dependent event

A dependent event has an outcome that is affected by the outcome of a previous event.

Example
You have only three coins in your bank. You know that they are a quarter, a dime, and a nickel. You shake out one coin, a quarter. You do not put it back. What coins are left to shake out next time?

In this situation, the second coin represents a dependent event, because the outcome depends on which coin you shook out the first time.

diagonal

For a polygon, a diagonal is a line segment joining two vertices that are not next to each other. One of the diagonals is shown for polygons A and B. All the diagonals are shown for polygons C and D.

 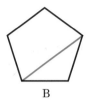 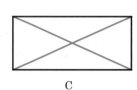 

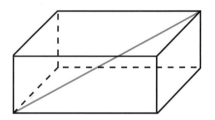

A   B   C   D

For a polyhedron, a diagonal is a line segment joining two vertices that are in different faces. One diagonal is shown for this polyhedron.

See also polygon, polyhedron.

diameter

The diameter of a circle is a line segment that passes through the center of the circle and has endpoints on the circle.

The diameter of a sphere is a line segment that passes through the center of the sphere and has endpoints on the sphere.

Circles and spheres have an infinite number of diameters.

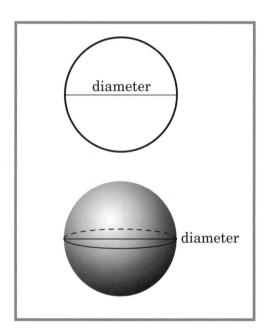

difference

The difference is the result of subtraction when two numbers are compared. It answers the questions
"How many more or fewer are there?" and
"How much more or less is there?"

Example

Martin wanted to buy a book on soccer. He found it for $19.95 at a local bookstore. He searched online and found it for $14.50. How much less did the book cost online?

19.95 − 14.50 = 5.45

The difference in the two prices was $5.45.

See also subtraction.

digit

A digit is a basic symbol used in a numeration system. The 10 digits used in our decimal, or base-ten, numeration system are

0, 1, 2, 3, 4, 5, 6, 7, 8, and 9.

?? Did You Know ??

Fingers and toes are also called digits.
Perhaps the reason for a numeration system
with 10 digits is that humans have 10 fingers!

dimension

The dimensions of some plane figures are length and width. The dimensions of some space figures are length, width, and height. The dimensions may also be the lengths of sides or edges. (A line segment has only one dimension: length.)

Example

The dimensions of this rectangular prism are

Length	4 centimeters
Width	2 centimeters
Height	1 centimeter

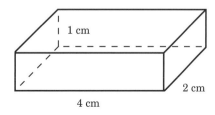

1 cm
2 cm
4 cm

See also one-dimensional, three-dimensional, two-dimensional.

discount

A discount is an amount by which the regular price of an item has been reduced. Discounts can be expressed as dollar amounts, common fractions, or percents.

Example

The discount on the bicycle, expressed in a dollar amount, is $25. It now costs $75. Expressed as a common fraction, the discount is $\frac{1}{4}$ ($25) off its regular price. It now costs $75. Expressed as a percentage, the discount is 25% ($25) off its regular price. It now costs $75.

distributive property of multiplication over addition

The distributive property of multiplication over addition can be expressed as
$$a(b + c) = (a \times b) + (a \times c).$$

Example

$$a(b + c) = (a \times b) + (a \times c)$$
$$3(4 + 2) = (3 \times 4) + (3 \times 2)$$

One use is to figure out harder multiplication facts. It uses easier multiplication facts.

Example

$3 \times 8 =$
$3 \times (4 + 4) =$
$(3 \times 4) + (3 \times 4) =$
$12 + 12 =$
24

$$3 \times 8 = (3 \times 4) + (3 \times 4)$$

$4 \times 9 =$
$4 \times (5 + 4)$
$(4 \times 5) + (4 \times 4) =$
$20 + 16 =$
36

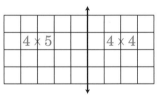

$$4 \times 9 = (4 \times 5) + (4 \times 4)$$

The distributive property of multiplication over addition also helps us multiply numbers with more than one digit.

Example

$$\begin{array}{r} 431 \\ \times\, 2 \\ \hline \end{array} \qquad \begin{array}{r} 400 \\ \times\, 2 \\ \hline 800 \end{array} \;+\; \begin{array}{r} 30 \\ \times\, 2 \\ \hline 60 \end{array} \;+\; \begin{array}{r} 1 \\ \times\, 2 \\ \hline 2 \end{array} \;=\; 862$$

distributive property of multiplication over addition
(continued)

It also helps us divide numbers.

Example

$$2\overline{)862} \quad 2\overline{)800} \quad + \quad 2\overline{)60} \quad + \quad 2\overline{)2} \quad = 431$$

with quotients 431, 400, 30, 1

divide

One number divides another number if there is 0 remainder after division.

Example

4 divides 12 because $12 \div 4 = 3$ with 0 remainder. Another way of saying that 4 divides 12 is to say that 12 is divisible by 4.

See also divisible.

dividend

A dividend is the number being divided.

Example

In $10 \div 5 = 2$,
10 is the dividend.
5 is the divisor.
2 is the quotient.

divisibility test

A divisibility test is a way of finding out if one number divides another number. Divisibility tests for 2, 3, 4, 5, 6, 9, and 10 are included here.

divisibility test for 2
If a number has an even number in the ones place,
2 divides the number.
For example, 2 divides 3,018 because 8 is an even number.

divisibility test for 3
If 3 divides the sum of the digits of a number,
3 divides the number.
For example, 3 divides 2310 because the sum of 2, 3, 1, and 0 is 6, and 3 divides 6.

divisibility test for 4
If 4 divides the last two digits of a number,
4 divides the number.
For example, 4 divides 8728 because 4 divides 28.

divisibility test
(continued)

divisibility test for 5
If a number has a 0 or 5 in the ones place,
5 divides the number.
For example, 5 divides both 385 and 400.

divisibility test for 6
If both 2 and 3 divide a number, 6 also divides that number.
For example, 6 divides 396 because both 2 and 3 divide 396.

divisibility test for 9
If 9 divides the sum of the digits of a number,
9 divides the number.
For example, 9 divides 198,432 because
$1 + 9 + 8 + 4 + 3 + 2 = 27$,
and 9 divides 27. So, 9 divides 198,432.

divisibility test for 10
If a number has 0 in the ones place, 10 divides it.
For example, 10 divides 710.

Also called divisibility rule.

divisible

A number is divisible by another number if the remainder is 0 after dividing.

Example
There are 32 children. Can they form 4 teams with no one being left out?

$$\begin{array}{r} 8 \text{ R0} \\ 4\overline{)32} \end{array}$$

They can form 4 teams with no one being left out because 32 is divisible by 4. (The remainder is 0 after dividing.)

divides
Another way of saying that 32 is divisible by 4 is to say that 4 divides 32. One number divides another number if the remainder is 0 after dividing.

division

Division is the inverse, or opposite, operation of multiplication. It "undoes" multiplication.

Example
$6 \times 4 = 24$, so $24 \div 4 = 6$.
In the example $24 \div 4 = 6$,
24 is the dividend.
4 is the divisor.
6 is the quotient.

division
(continued)

Written with symbols:
Where c is the dividend, b is the divisor, and a is the quotient, $c \div b = a$ because $a \times b = c$.

division by zero
Note that the divisor b cannot be zero because there is no number a that will multiply b to equal c when $c \neq$ zero.

sharing
Division can be used to find the size of a part of a set. This is called sharing, or partition division. When division is used to find the size of a part, a number is divided into equal parts.

There are 12 cookies. They are divided equally among 4 people. How many cookies does each person get?
    $12 \div 4 = 3$
Each person gets 3 cookies. (The size of each person's part, or share, is 3.)

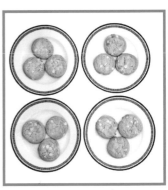

measurement division
Division is also used to find the number of parts in a set. This is called measurement division.

There are 12 cookies. We want to give 4 cookies to each person. How many people will get cookies?
    $12 \div 4 = 3$
3 people get cookies. (The set of cookies is divided into 3 parts.)

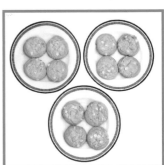

See also dividend, divisor, quotient.
Related words divide, dividend, divisible, divisive.

division sentence

A division sentence is a number sentence used to express division, in which the numbers are usually written the following order:
Dividend ÷ divisor = quotient.

Example
    $15 \div 3 = 5$
This division sentence can also be written another way using a fraction.
    $\frac{15}{3} = 5$

See also number sentence.

divisor

A divisor is the number by which another number is divided.

Example
In $14 \div 2 = 7$,
14 is the dividend.
2 is the divisor.
7 is the quotient.

fig. 1

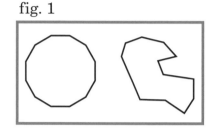

dodecagon

A dodecagon is a polygon with 12 sides (fig.1).
(*Do-* means "two" and *deca-* means "ten.")

fig. 2

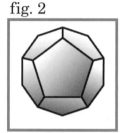

dodeca-hedron

A dodecahedron is a space figure with 12 faces (fig. 2). If each is in the shape of a regular pentagon, it is one of the five regular polyhedra.

See also polyhedron.

fig. 3

double

A double is an addition fact in which the two addends are the same number (fig. 3).

Example
$5 + 5 = 10$

Table of Basic Addition Facts

+	0	1	2	3	4	5	6	7	8	9
0	0	1	2	3	4	5	6	7	8	9
1	1	2	3	4	5	6	7	8	9	10
2	2	3	4	5	6	7	8	9	10	11
3	3	4	5	6	7	8	9	10	11	12
4	4	5	6	7	8	9	10	11	12	13
5	5	6	7	8	9	10	11	12	13	14
6	6	7	8	9	10	11	12	13	14	15
7	7	8	9	10	11	12	13	14	15	16
8	8	9	10	11	12	13	14	15	16	17
9	9	10	11	12	13	14	15	16	17	18

double bar graph

See bar graph.

double line graph

See line graph.

dozen

A dozen is a set of 12 items.
A dozen eggs are pictured here.

In olden days, bakers would often give their customers an extra item when they bought 12, "for good measure." Hence the term "baker's dozen" for 13 items.

dry measure   Dry measure is used to measure the volume of items such as fruits, vegetables, and grains. In the system of dry measure:

2 pints = 1 quart
8 quarts = 1 peck
4 pecks = 1 bushel

Dry pints and dry quarts are a little larger than liquid pints and liquid quarts.

See also dry pint, dry quart.

dry pint   A dry pint is a unit of capacity in the customary system of measurement. It is slightly larger than a liquid pint and is used to measure dry products such as fruits and grains.

See also dry measure, pint.

dry quart   A dry quart is a unit of capacity in the customary system of measurement. It is slightly larger than a liquid quart and is used to measure dry products such as fruits and grains.

See also dry measure, quart.

Ee

edge

An edge is a line segment formed where two faces of a space figure meet.

A square pyramid has 8 edges.

A cube has a total of 12 edges.

This picture of a cliff provides a model for an edge.

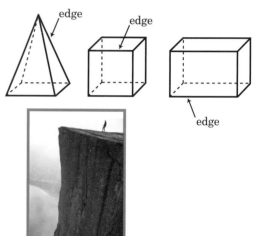

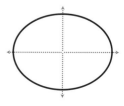

elapsed time

Elapsed time is the amount of time that has passed.

Example

Lunch is served at 11:30 AM. Dinner is at 6:00 PM. The elapsed time between lunch and dinner is $6\frac{1}{2}$ hours.

element

An element is a member of a set.

Example

The set of whole numbers is {0, 1, 2, 3, 4, 5, 6, ... }. 4 is one of the elements of the set of whole numbers.

See also set.

ellipse

An ellipse is a special kind of oval that is mathematically regular, with two lines of symmetry. We usually think of it as looking like a "flattened" or "stretched" circle.

ellipse
(continued)

Mathematically, a circle is a special example of an ellipse, just as a square is a special example of a rectangle.

?? Did You Know ??

The paths (orbits) of the planets around the sun are elliptical.

See also oval.
Related word elliptical.

empty set

An empty set is a set with no members.

Example

The set of humans 20 feet tall is an empty set.

See also zero.

endpoint

An endpoint is a point at the end of a line segment or ray.

Points A and B are endpoints for the line segment.

The ray has only one endpoint: Point C.

Line Segment AB   Ray CD

equal

Equal means having the same amount, size, or value or being identical.

Examples

There is an equal number of cups and saucers.

The pizza is cut into pieces of equal size.

12 inches are equal to 1 foot.

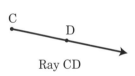

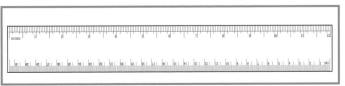

equal
(continued)

The area of this rectangle is equal to the product of its length and width.

$$A = l \times w$$

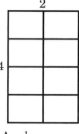

$A = l \times w$

$A = 4 \times 2$

$A = 8$ square units

See also equal sign, equality, equivalent.
Related words equality, equation, inequality.

equality

Equality is a relationship between two quantities that are of the same amount, size, or value. When an equality is a number relationship, it is expressed by a mathematical sentence that uses the equal sign (=).

Examples

$4 + 6 = 10$ Both $4 + 6$ and 10 name the same number.

$3n = 12$ In this equation, $3n$ and 12 name the same number.

See also equation.

equally likely outcomes

Equally likely outcomes are outcomes that have the same chance of occurring.

See also outcome.

equal ratios

Equal ratios are ratios that describe the same rate or make the same comparison.

Example

Samantha wants to buy and stock an aquarium.
She saves $1 of each $5 she earns.

Saved	$1	$2	$3	$4	$5
Earned	$5	$10	$15	$20	$25

1:5, 2:10, 3:15, 4:20, and 5:25 are all equal ratios.

These ratios can also be written as $\frac{1}{5}$, $\frac{2}{10}$, $\frac{3}{15}$, $\frac{4}{20}$, and $\frac{5}{25}$.

See also proportion, ratio.

equal sign (=)

Read as *is equal to* or *equals*.
The equal sign (=) indicates
that one amount, size, or value is the same as another.

Examples

$2 + 3 = 5$
1 hour = 60 minutes
$\frac{1}{4} = \frac{2}{8}$

equation

An equation is a mathematical sentence that gives
two names for the same number. It is written with
an equal sign (=).

variable
An equation can have missing numbers. Symbols that represent
the missing numbers are called variables.

Examples

$7 + \square = 63$ $\square$ is the variable
$y \times 8 = 48$ y is the variable

See also variable.

equilateral

Equilateral means having all sides of equal length.

equilateral polygon

An equilateral polygon has all sides of equal length.

See also equilateral, polygon.

equilateral triangle

An equilateral triangle has all sides of
equal length.

See also equilateral, triangle.

?? Did You Know ??

The Star of David is a symbol of the Republic of Israel. It is formed by two interlocking equilateral triangles. You can find it on Israel's flag.

equivalent

Equivalent is often used to mean having the same value.

Examples

5 pennies are equivalent to 1 nickel.

$\frac{4}{4}$ is equivalent to $\frac{2}{2}$ or 1.

$4 + 2$ and $7 - 1$ are equivalent expressions. They name the same number.

See also equal.

equivalent fractions

Equivalent fractions are fractions that name the same number. The fractions $\frac{2}{2}$ and $\frac{4}{4}$ name the same number, 1.

To name equivalent fractions, multiply the fraction by a fractional name for 1 ($\frac{2}{2}$, $\frac{3}{3}$, $\frac{4}{4}$, $\frac{5}{5}$, ...).

Examples

$\frac{1}{2} \times \boxed{\frac{2}{2}} = \frac{2}{4}$ $\frac{1}{2} \times \boxed{\frac{3}{3}} = \frac{3}{6}$ $\frac{1}{2} \times \boxed{\frac{4}{4}} = \frac{4}{8}$

Equivalent fractions are used when adding and subtracting unlike fractions.

Example

$\frac{1}{4} + \frac{1}{3} = \boxed{}$

First: Find a common multiple of 4 and 3. Since 12 is a common multiple, it can be used as a denominator for both fractions.

Next: Express $\frac{1}{4}$ and $\frac{1}{3}$ as equivalent fractions that have a denominator of 12.

$$\frac{1}{4} \times \boxed{\frac{3}{3}} = \frac{3}{12}$$
$$\frac{1}{3} \times \boxed{\frac{4}{4}} = \frac{4}{12}$$
$$\frac{3}{12} + \frac{4}{12} = \frac{7}{12}$$

Also called equal fractions.

See also equal.

estimate

An estimate is a number close to an exact number. It is used when an exact number cannot be found or is not needed. To estimate is to find such a number.

See also estimation strategies.
Related word estimation.

estimation strategies

Estimation strategies are ways used to find a number that is close enough to an exact number for our purposes.

amounts

Often we want to estimate how much or how many there are. In these cases the use of a benchmark is a helpful strategy. (A benchmark is a reference point that can be used to help in making an estimate.)

Example

Question: Do we have enough milk for 4 people to have cereal? (We usually use a little less than a quart of milk.)

Benchmark: We know that 1 quart = $\frac{1}{4}$ gallon. If the container is filled about $\frac{1}{4}$ of the way, it contains about 1 quart.

Answer: The gallon container of milk is more than $\frac{1}{4}$ full, so there is plenty for breakfast cereal.

computational estimation

We use computational estimation when we want to predict what the result of an operation will be. We do not need to do the computation to get an exact answer when an approximate number is close enough for our purposes.

Five major strategies for computational estimation are clustering, compatible numbers, front-end estimation, rounding, and finding a range.

clustering

Clustering is a method used for estimating a result when numbers appear to group, or cluster, around a common number.

**estimation
strategies**
(continued)

Example

Kathy sold 32 boxes of Girl Scout cookies on Monday, 25 boxes on Tuesday, and 36 boxes on Wednesday. About how many boxes of cookies did she sell in 3 days?

   32, 25, and 36 cluster around 30.
   $3 \times 30 = 90$

Kathy sold about 90 boxes of cookies in 3 days.

compatible numbers
Compatible numbers are numbers that seem to "go together." They are easy to compute mentally.

Example

Robert bought 3 pencils for $0.79. How much did he pay for each pencil?

   $0.79 is close to $0.75, a number compatible with 3.
   (Three quarters are equal to $0.75.)

Robert paid a little more than a quarter ($0.25) for each pencil.

front-end estimation
Front-end estimation is a strategy for estimating a result by using the first digit at the front of numbers, followed by zeros, in computing.

Example

Brent and his friends went to the ice-cream shop after the ball game. They bought 5 milk shakes for $2.09 each. About how much did they spend?

   Front-end $2.09 to $2.00.
   $5 \times \$2.00 = \$10.00$

Brent and his friends spent about $10.00 for milk shakes.

rounding
Rounding strategies are sometimes used in computational estimation. Rounding a whole number results in a number close to the original number. It usually has the same number of digits, but more of them are zeros.

Example

Salita's mother commutes from New York to Washington, DC, once each week. The round-trip flight is about 370 miles. About how many miles does she commute each month?

   370, rounded to the nearest 100, is 400.
   There are about 4 weeks in a month.
   $4 \times 400 = 1,600$

Salita's mother commutes about 1,600 miles each month.

estimation strategies
(continued)

finding a range

Finding a range also uses rounding for computational estimation. The numbers being used are rounded up to make an estimate. This estimate gives the upper end of the range within which the exact answer falls. The numbers being used are also rounded down to make an estimate. This estimate gives the lower end of the range within which the exact answer falls. The exact answer will fall within the range between the two estimates.

Example

Tammy and Clinton were camping in Yellowstone National Park. They wanted to travel south to Jackson, Wyoming, and then on to Bear Lake State Park in Utah.
About how many miles did they need to travel?

Yellowstone to Jackson	98 miles
Jackson to Bear Lake State Park	146 miles

Estimate 1

> 98, rounded up, is 100.
> 146, rounded up, is 150.
> 100 + 150 = 250
> 250 is the upper end of the range.

Estimate 2

> 98, rounded down, is 90
> 146, rounded down, is 140.
> 90 + 140 = 230
> 230 is the lower end of the range.
> 98 + 146 is in the range between 230 and 250.

Tammy and Clinton need to travel between 230 and 250 miles from Yellowstone to Bear Lake State Park.

See also clustering, compatible numbers, front-end estimation, rounding.

even number

An even number is a whole number that can be divided by 2 with 0 remaining.

Example

Even numbers have 0, 2, 4, or 8 in the ones place.

Gloves, socks, and shoes are packaged in even numbers. So are most other items that have more than one per package.

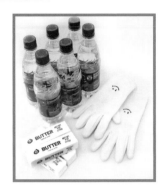

event

Events are the possible outcomes of a probability experiment.

Example
If you roll a die, some of the events include rolling a number greater than 4, rolling a 6, rolling a number less than or equal to 3, and rolling a 4.

expanded form

The expanded form of a number is that number written in a way that shows the value of each of its digits.

There are several ways to write a number in expanded form.

Examples (for 423)
$$400 + 20 + 3$$
$$(4 \times 100) + (2 \times 10) + (3 \times 1)$$
$$(4 \times 10^2) + (2 \times 10^1) + (3 \times 10^0)$$
$$(4 \times 10^2) + (2 \times 10) + (3 \times 1)$$

Also called expanded notation. When exponents are used, ***also called*** exponential notation.

experiment

In probability, an experiment is an activity that has two or more possible results or outcomes. Some simple probability experiments include tossing a coin and rolling a die.

exponent

An exponent is a number that tells how many times the base is used as a factor.

exponent
(continued)

Example

In 5^3, 5 is the base and 3 is the exponent.
($5^3 = 5 \times 5 \times 5 = 125$)

expression

An expression is a mathematical phrase without an equal sign.

Examples

If one pair of shoes costs $19.99, how can the cost of 4 pairs of shoes at the same price be expressed? $4 \times \$19.99$ is an expression for the cost of 4 pairs of shoes.

In algebra, expressions include variables and operations.
$3 + x$ is an expression that means 3 plus a number.
$5n$ is an expression that means the product of 5 and a number.
$\frac{y}{2}$ means a number divided by 2.

See also variable.

face

A face can be any polygon used in forming a space figure. Not all space figures have faces. For instance, a pyramid has faces but a sphere does not.

polyhedron

A space figure with polygons for faces is a polyhedron.

?? Did You Know ??

This picture of a rock climber shows the rock face he is climbing. How is the rock face like the face of a space figure?

See also base, polyhedron.

fact family

A fact family is a set of related addition and subtraction facts or multiplication and division facts.

An addition and subtraction fact family uses two addends and their sum to form basic facts.

Examples

Fact family using 3, 4, and 7	Fact family using 8, 0, and 8
3 + 4 = 7	8 + 0 = 8
4 + 3 = 7	0 + 8 = 8
7 − 4 = 3	8 − 0 = 8
7 − 3 = 4	8 − 8 = 0

fact family
(continued)

A multiplication and division fact family uses two factors and their product to form basic facts.

Examples

Fact family using 4, 6, and 24	Fact family using 5, 0, and 0
$4 \times 6 = 24$	$5 \times 0 = 0$
$6 \times 4 = 24$	$0 \times 5 = 0$
$24 \div 6 = 4$	$0 \div 5 = 0$
$24 \div 4 = 6$	(There is no fourth member of this fact family. $5 \div 0$ is undefined.)

Also called related facts, related sentences.

See also basic facts, related facts.

factor

Factors are any of the numbers multiplied to form a product.

Example

In the multiplication sentence $4 \times 3 = 12$, 4 and 3 are factors and 12 is the product. Both 4 and 3 divide 12. That is, $12 \div 4 = 3$ with 0 remaining, and $12 \div 3 = 4$ with 0 remaining.

factors of a number

The factors of a number are all the numbers that divide that number.

Example

The whole number factors of 12 are 1, 2, 3, 4, 6, and 12.

factoring

Factoring is the process of writing a number or expression as the product of its factors.

See also common factor, factor pair, factor tree, greatest common factor, prime factor, prime factorization, proper factor.

factor pair

A factor pair is a set of two numbers that, when multiplied, will result in a given product.

Example

In $2 \times 6 = 12$, the factor pair for 12 is (2, 6).
Other factor pairs for 12 include (1, 12) and (3, 4).

factor tree

A factor tree is a diagram showing factors of a number. The diagram is complete when the prime factors of that number have been found. Different factor trees may be used for the same number, but the prime factorization is the same.

Example

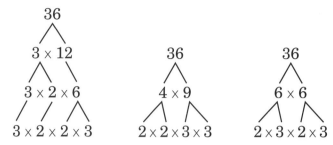

The prime factorization of 36 is $2 \times 2 \times 3 \times 3$ (or $2^2 \times 3^2$).

See also prime factor, prime factorization.

Fahrenheit temperature scale (°F)

The Fahrenheit (°F) temperature scale is used for measuring temperature in the customary system. The temperature is expressed in *degrees Fahrenheit (°F)*.

Two reference points for this temperature scale are

32° (the freezing point of water)
212° (the boiling point of water)

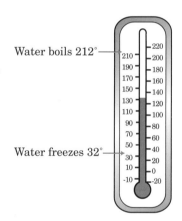

Water boils 212°

Water freezes 32°

?? Did You Know ??

The Fahrenheit temperature scale was developed in the early 1700s by a German scientist named Gabriel David Fahrenheit. He was the first person to make a mercury thermometer.

Also called Fahrenheit scale.

See also Celsius (°C) temperature scale, temperature.

fathom

A fathom is a unit of length equal to 6 feet. It is used for measuring the depth of water.

fathom
(continued)

?? Did You Know ??

Samuel Langhorne Clemens was one of the most famous Americans of the 1800s. The author of several books, including *The Adventures of Tom Sawyer*, he is best remembered by his pen name, Mark Twain. This name may have come from his work as a steamboat pilot. To steer the steamboat, the water needed to be 2 fathoms deep, or "by the mark, twain" to be in safe water.

favorable outcome

A favorable outcome is a result that meets the condition being investigated in an experiment.

Example
When a coin is tossed to see how likely it is to land "tails," a favorable outcome is landing on its "tail."

See also outcome.

figurate numbers

Figurate numbers are numbers that can be represented by arrays that look like various plane figures.

Also called polygonal numbers.
See also pentagonal numbers, polygonal numbers, square numbers, triangular numbers.

flip

A flip is a mirror image of a figure.

Also called reflection.
See also slide, turn.

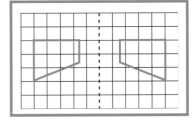

fluid ounce (fl oz)

A fluid ounce is a unit of capacity in the customary system of measurement.

8 fluid ounces = 1 cup

See also capacity, customary system of measurement, ounce.

foot

plural **feet**

A foot is a unit of length in the customary system of measurement.

     1 foot = 12 inches

See also customary system of measurement.

formula

A formula is an equation that expresses a mathematical relationship, principle, or rule.

Example

Makenna and her family were planning a long road trip. They planned to travel 8 hours a day by car, averaging about 55 miles each hour. What is the distance they could travel each day?

This problem can be solved using the following formula.

$D = R \times T$
(Distance is equal to the Rate of travel times the Time traveled.)
$D = 55 \times 8$
$D = 440$, the number of miles that Makenna's family could travel each day.

fraction

A fraction is a number that can be expressed as $\frac{a}{b}$, in which a and b are any number and b is not equal to $\frac{a}{b}$.

Fraction is often used as a shortened form of the term *common fraction*, but fractional numbers can also be expressed as decimal fractions and percents.

See also common fraction, decimal fraction, fractional number, percent, rational number.

fractional number

A fractional number is used to express a part of a whole or a group. A fractional number may also be used to express a ratio. A fractional number may be written in three different ways: As a common fraction, a decimal fraction, or a percent.

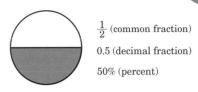

fractional number
(continued)

Examples

$\frac{1}{2}$, 0.5, or 50% of this circle is shaded.

$\frac{1}{2}$ (common fraction)

0.5 (decimal fraction)

50% (percent)

$\frac{1}{2}$, 0.5, or 50% of these squares are shaded.

$\frac{1}{2}$ (common fraction)

0.5 (decimal fraction)

50% (percent)

$\frac{1}{2}$, 0.5, and 50% are different names for the same fractional number.

See also common fraction, decimal fraction, fractional number, percent, rational number.

fraction bar

A fraction bar is a horizontal bar that separates the numerator and the denominator of a common fraction. The numerator is written above the fraction bar. The denominator is written below the fraction bar.

$$\frac{3}{4}$$

← numerator
← fraction bar
← denominator

frequency

Frequency is the number of times an event occurs within a specific time period.

Example

The frequency of earthquakes at a magnitude of 8.0 or higher on the Richter Scale is an average of 1 per year worldwide, according to the National Earthquake Information Center.

Frequency can also mean the number of times an item occurs in a set of data.

Example

In our family, 2 people have birthdays on January 23. The frequency of birthdays on January 23 is 2.

See also frequency table.

frequency table

A frequency table is used to summarize the number of times items occur in a set of data. Tally marks are often used to record the number of times, or the frequency, of an item.

How We Get to School	
Walk	卌 ‖‖
Ride a bus	‖‖‖
Ride in a car	‖
Ride a bicycle	卌 卌 ‖

Example

In the frequency table to the right, the frequency for riding a bicycle to school is 11.

Also called tally chart.

front-end digit

The front-end digit in a number is the first digit, reading the number from left to right. It is the digit in the number that has the greatest place value.

See also front-end estimation.

front-end estimation

Front-end estimation is a strategy for estimating sums, differences, products, or quotients by using front-end digits followed by zeros. To do front-end estimation, keep the first digit at the front of each number to be used (the front-end digit). Use zeros in place of the remaining digits. Then add, subtract, multiply, or divide mentally.

Example

Population estimates for the United States in 2003 showed that California had the most people (35,484,453), with Texas in second place (22,118,509). About how many more people lived in California than in Texas?

$$35,484,453 \rightarrow 35,000,000$$
$$22,118,509 \rightarrow 22,000,000$$
$$35,000,000 - 22,000,000 = 13,000,000$$

About 13,000,000 more people lived in California than in Texas in 2003.

See also estimation strategies.

ft²

Read as *square foot.*

A ft² is the amount of area enclosed by a square that measures 1 foot by 1 foot.

Also written as square foot and sq ft.
See also square unit.

ft³

Read as *cubic foot.*

A ft³ is equal to the volume of a cube that measures 1 foot on each edge.

Also written as cubic foot and cu ft.
See also cubic unit.

function

A function is a relation between two sets in which each member of the first set is paired with one and only one member of the second set.

Example

Two of the ways that functions may be shown are by using tables (such as the one at right) and by mapping (shown below).

Function Table	
A	B
0	0
12	3
28	7
36	9
32	8
24	6
8	2
20	5
4	1
16	4

Function Rule:
A is the product of B and 4

Mapping: Creatures and Sounds

C (Creatures)   S (Sounds)

Bird → buzz
Bee → roar
Cat → meow
Dog → chirp
Lion → bark

Function Rule: C is the animal that makes the sound S.

Also called mapping.

function rule

A function rule is a rule that explains the relationship between two sets.

See also function.

furlong

A furlong is a unit of length equal to $\frac{1}{8}$ mile. It is often used to measure distances of horse races.

furlong
(continued)

The Kentucky Derby, the world's most famous horse race, is held on the first Saturday of May each year at the Churchill Downs in Louisville. It is a 10-furlong race ($1\frac{1}{4}$ miles) for 3-year-old horses.

gallon

A gallon is a unit of capacity in the customary system of measurement.

1 gallon = 4 quarts

?? Did You Know ??

A half-gallon carton of ice cream will serve a cup of ice cream to 8 people.

See also capacity, customary system of measurement.

GCF (greatest common factor)

The GCF is the largest factor that two or more numbers share.

See also greatest common factor.

geobands

Geobands are rubber bands used for constructing geometric figures on geoboards.

See also geoboard.

geoboard

A geoboard is a board with regularly spaced pegs. It is used for exploring plane figures.

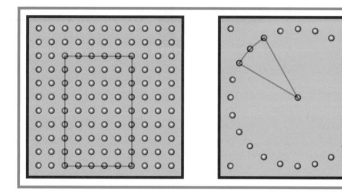

geometric figure

A geometric figure is any combination of points, lines, or planes.

space figure
A space figure is a three-dimensional geometric figure, or a figure that occupies space and has volume.

Example

This tent is in the shape of a space figure, a triangular prism.

This drawing shows the three dimensions of a triangular prism.

plane figure
A plane figure is a two-dimensional geometric figure. It has no thickness and lies entirely in one plane.

Example

This triangular prism has plane figures for faces. Two of the faces are triangles.

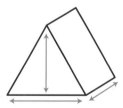

lines, line segments, and rays
Lines, line segments, and rays are one-dimensional geometric figures.

geometric figure
(continued)

Example
The base of this triangle is a line segment.
It has one dimension, length.

points
Look at the triangle above and locate the vertices, labeled A, B, and C. These points could be called zero-dimensional geometric figures because they do not occupy space.

Also called geometric shape.

See also line, line segment, one-dimensional, plane figure, point, ray, space figure, three-dimensional, two-dimensional, zero-dimensional.

geometry

Geometry is a branch of mathematics that includes the study of shape, size, and other properties of figures. It is one of the oldest branches of mathematics, probably used even in prehistoric times.

?? Did You Know ??

The kind of geometry most students usually study was recorded in *The Elements*, a set of books written about 300 BC by Euclid, a Greek mathematician. There are now other kinds of geometry as well. These other kinds originated about 2,000 years after Euclid's work.

Related word geometric.

Golden Rectangle

A Golden Rectangle is a rectangle for which the ratio of length to width is approximately 8 to 5. Many famous buildings have a face in the shape of a golden rectangle.

See also rectangle.

googol

A googol is a number equal to 1 followed by 100 zeros, or 10^{100}.

googol
(continued)

In the late 1930s, Edward Kasner, an American mathematician, was working with a very large number. He asked his 9-year-old nephew what he might call this number, written as 10^{100}. His nephew named it *googol*.

Related word googolplex.

gram (g)

A gram is a unit of weight in the metric system of measurement.

1,000 grams = 1 kilogram

See also metric system of measurement.

graph

A graph is a kind of drawing that shows mathematical information, ideas, and relationships. In statistics, examples of graphs used to represent data are bar graphs, picture graphs, circle graphs, line graphs, line plots, and stem-and-leaf plots.

In algebra and geometry, number lines are used to graph numbers. The coordinate system is used to graph ordered pairs, equations, and other mathematical relationships.

See also bar graph, circle graph, coordinate system, double bar graph, line graph, line plot, picture graph, real graph, stem-and-leaf plot.

graph scale

Graph scale is the ratio between the picture, or icon, on a graph and the number it represents.

See also graph, scale.

greater than or equal to sign (≥)

Read as *is greater than or equal to*. The greater than or equal to sign (≥) is used to compare two numbers when the first number expressed is greater than or equal to the second number.

Example

Federal law requires that the weight of a product must be greater than or equal to the amount listed on the label. For

greater than or equal to sign (≥)
(continued)

instance, a bag of potatoes labeled as weighing 10 pounds must weigh *at least* 10 pounds.

bag of potatoes ≥ 10 lb

See also inequality.

greater than sign (>)

Read as *is greater than*. The greater than sign (>) is used to compare two numbers when the greater number is expressed first.

Examples

8 > 7   Read as *eight is greater than seven.*
$2 \times 3 > 2 \times 2$ Read as *two times three is greater than two times two.*

See also inequality.

greatest common factor (GCF)

The greatest common factor (GCF) of two or more numbers is the largest factor they share. The greatest common factor may be found by listing factors or by using prime factorization.

For smaller numbers, listing factors is usually not too hard to do and is helpful in finding the greatest common factor.

Example

To find the greatest common factor of 8 and 12:

First: list the factors.

Factors of 8 are 1, 2, 4, 8.
Factors of 12 are 1, 2, 3, 4, 6, 12.

Next: compare these factors and identify the largest factor that the two numbers share.

The greatest common factor of 8 and 12 is 4.
This can also be written as GCF (8, 12) = 4.

For larger numbers, listing factors is usually more difficult. In these cases, prime factorization is helpful.

greatest common factor (GCF)

(continued)

Example

To find the greatest common factor of 36 and 48:
First: find the prime factorization.

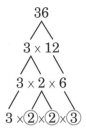 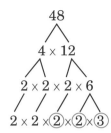

The numbers 36 and 48 have the prime factors 2, 2, and 3 in common.

Next: multiply these prime factors: $2 \times 2 \times 3 = 12$
GCF (36, 48) = 12

Knowing how to find the greatest common factor is helpful in expressing common fractions in their simplest form.

Also called greatest common divisor.

See also common factor, prime factorization, simplest form.

grid

A grid is a set of uniformly spaced horizontal and vertical line segments.

See also coordinate system.

gross

A gross is equal to 12 dozen (12×12), or 144. Gross also is the total before any deductions.

Example

The gross income from the bake sale was $75.

?? Did You Know ??

School pencils are often ordered by the gross. There are 12 pencils in each package, and 12 packages in each gross of pencils.

See also gross income.

gross income

Gross income is the total amount of money earned before taxes and other items are deducted.

grouping property

The grouping property means that changing the grouping of the numbers used in an operation without changing the order does not change the result of that operation. Addition and multiplication have the grouping property, but subtraction and division do not.

Also called associative property.
See also associative property.

grouping property of addition

The grouping property of addition means that when adding three or more numbers, we can group the numbers in any way we choose without changing the order and the sum will remain the same.

Also called associative property of addition.
See also associative property of addition.

grouping property of multiplication

The grouping property of multiplication means that when multiplying three or more numbers, we can group the numbers in any way we choose without changing the order and the product will remain the same.

Also called associative property of multiplication.
See also associative property of multiplication.

hecto-

Hecto- is a prefix meaning hundred.

Examples

> 1 hectometer = 100 meters
> 1 hectogram = 100 grams
> 1 hectoliter = 100 liters

See also metric system of measurement.

hectogram (hg)

A hectogram is a unit of weight in the metric system of measurement.

> 1 hectogram = 100 grams

See also metric system of measurement.

hectoliter (hL)

A hectoliter is a unit of capacity in the metric system of measurement.

> 1 hectoliter = 100 liters

See also metric system of measurement.

hectometer (hm)

A hectometer is a unit of length in the metric system of measurement.

> 1 hectometer = 100 meters

?? Did You Know ??

The Statue of Liberty, one of the world's tallest statues, measures about 93 meters from its foundation to the top of the torch. This distance is nearly one hectometer.

See also metric system of measurement.

height

Height refers to how tall somebody or something is. The height of a geometric figure is its altitude.

> ### *Example*
>> The tallest animal is the giraffe.
>> It grows to a height of about 5.5 meters.
>>
>> The tallest living thing is believed to be a redwood tree, reaching a height of about 105 meters.

See also altitude.

hemisphere

A hemisphere is half of a sphere. (*Hemi-* means half.)

Earth can be divided into the Northern Hemisphere and the Southern Hemisphere.

Earth can also be divided into the Eastern Hemisphere and the Western Hemisphere.

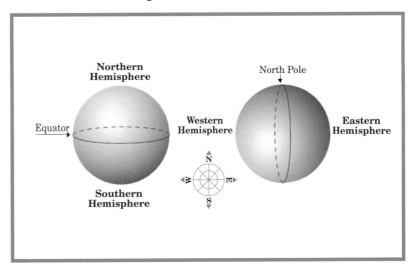

heptagon

A heptagon is a polygon with 7 sides. (*Hepta-* means seven.)

See also polygon.

Regular heptagon   Not regular heptagons

hexagon

A hexagon is a polygon with 6 sides (*Hexa-* means six.)

See also polygon.

Regular hexagon   Not regular hexagons

Hindu-Arabic numeration system

The Hindu-Arabic numeration system is a base-ten system of numeration in which the digits 0, 1, 2, 3, 4, 5, 6, 7, 8, and 9 are used.

See also decimal numeration system.

horizontal axis

The horizontal axis is the horizontal number line in a rectangular coordinate system.

Also called x-axis.

See also coordinate system, x-axis.

horizontal bar graph

A horizontal bar graph is a kind of graph that uses rectangular bars to show information. The bars go from left to right.

See also bar graph.

hour (hr)

An hour is a unit of time. It is used in both the customary and metric systems of measurement and is abbreviated as *hr*.

> 60 minutes = 1 hour
> 24 hours = 1 day

Hour can also mean a set or appointed time.

Examples

> Paula's lunch hour is at 11:30 AM.
> The plane departs at 2100 hours (military time).

hundred

A hundred is equal to 10 tens or 100 ones. In standard form, one hundred is written as 100.

Using an exponent, 100 may be written as 1×10^2 (or simply as 10^2).

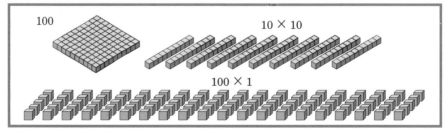

See also decimal numeration system.

hundredth

One hundredth is one of 100 equal parts of a whole or a group. One hundredth may be written as $\frac{1}{100}$ or 0.01.

Examples

One penny or cent is equal to one hundredth of a dollar.

The unit, or cube, is one hundredth of the flat.

In ordinal numbers, hundredth is next after ninety-ninth.

See also decimal numeration system, ordinal number.

hundredths

In the decimal numeration system, hundredths is the name of the next place to the right of tenths. In the number 2.98, 8 is in the hundredths place.

See also decimal numeration system, hundredth.

ones	decimal point	tenths	hundreths
2	.	9	8

icosahedron

An icosahedron is a space figure with 20 faces.

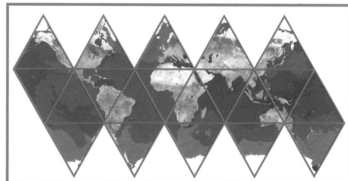

You can create a three-dimensional version of this Earth icosahedron by following the instructions on page 200 of this book.

See also polyhedron.

identity element

An identity element is a number that, when used in an operation with another number, leaves that number the same.

The identity element for addition is 0. When 0 is added to any number, that number remains the same, or keeps its identity.

Example

$4 + 0 = 0 + 4 = 4$

The identity element for multiplication is 1. When 1 is multiplied by any number, that number remains the same, or keeps its identity.

Example

$1 \times 4 = 4 \times 1 = 4$

See also identity property for addition, identity property for multiplication.

identity property for addition

The identity property for addition means that the sum of 0 and any number is that number.

Example

27 + 0 = 0 + 27 = 27

3 + 0 = 0 + 3 = 3

The identity property for addition may be written with symbols as

$a + 0 = 0 + a = a$

Also called identity property of 0 for addition, zero property of addition.

See also identity element, identity property for multiplication.

identity property for multiplication

The identity property for multiplication means that the product of 1 and any number is that number.

Example

$8 \times 1 = 1 \times 8 = 8$

The identity property for multiplication may be written with symbols as

$a \times 1 = 1 \times a = a$

Also called identity property of 1 for multiplication.

See also identity element, identity property for addition.

improper fraction

An improper fraction is a common fraction that names a number equal to or greater than 1.

Examples

$\frac{3}{2}$ $\frac{8}{4}$ $\frac{19}{6}$ $\frac{4}{4}$ $\frac{100}{75}$

For each improper fraction, the numerator is equal to or greater than the denominator.

See also common fraction, proper fraction.

in.²

Read as *square inch*.

A in.² is the amount of area enclosed by a square that measures 1 inch by 1 inch.

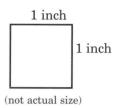

1 inch

1 inch

(not actual size)

Also written as square inch, sq in.
See also square unit.

in.³

Read as *cubic inch*.

A in.³ is equal to the volume of a cube that measures 1 inch on each edge.

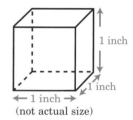

1 inch

1 inch

1 inch

(not actual size)

Also written as cubic inch.
See also cubic unit.

inch (in.)

An inch is a unit of length in the customary system of measurement.

12 inches = 1 foot

?? Did You Know ??

At one time in English history, an inch was defined as the width of a man's thumb. Today an inch can be defined as 2.54 cm, which is about the length of a child's thumb from the tip to the first joint.

See also customary system of measurement.

independent events

Independent events are events that have no effect on each other. Using a computer to do my homework tonight probably has no effect on what I have for breakfast tomorrow morning.

See also compound event, dependent event, event.

inequality

An inequality is a relationship between two different quantities. An inequality may be expressed by a mathematical sentence that uses one of the following symbols.

< is less than
> is greater than
≤ is less than or equal to
≤ is greater than or equal to
≠ is not equal to

Examples

$\frac{1}{2} < \frac{3}{4}$ $3 \times 2 \leq 4 + 3$
$99.8 > 98.6$ $11 \geq 9$
$2 + 3 \neq 2 \times 3$

See also equality, equal sign (=), greater than or equal to sign (≥), greater than sign (>), less than or equal to sign (≤), less than sign (<), not equal to sign (≠).

infinite

Infinite means unending. The symbol for infinity looks similar to a sideways figure 8.

Example

The set of even numbers is infinite: 2, 4, 6, 8, 10, 12, …
The three dots (…) are used to show that the set is unending.

Related word infinity.

integers

Integers are the counting numbers, their opposites, and zero.

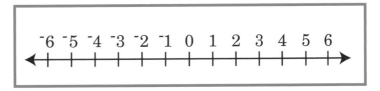

positive integers
Positive integers are integers greater than 0.

negative integers
Negative integers are integers less than 0.
0 is neither positive nor negative.

interest

Interest is the charge for borrowing money or the amount paid for money invested.

simple interest

To find simple interest, multiply the amount of money (principal) by the percent (rate) by the number of years (time).

Expressed as a formula:

Interest = principal × rate × time
or $I = p \times r \times t$

Example

Becky borrowed $20 from her mother for six months at 6% interest. How much did she owe her mother at the end of six months?

$I = p \times r \times t$
p (principal) = $20
r (rate) = 6%, or 0.06
t (time) = 6 months, or $\frac{1}{2}$ year
$I = \$20 \times 0.06 \times \frac{1}{2} = \$0.60$

At the end of six months, Becky owed her mother the principal, $20, plus the interest, $0.60, for a total of $20.60.

See also principal, rate.

intersect

Intersect means to meet, cross, or overlap.

lines, rays, line segments

For lines, rays, and line segments, intersect means to meet or cross. When two lines, rays, or line segments intersect, they have one common point.

Examples

This windmill's blades represent line segments that intersect, or cross.

Point (3, 4) is the intersection of line $x = 3$ and line $y = 4$. The point can be located by following the red arrows on the grid.

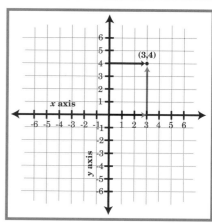

intersect
(continued)

other geometric figures

For other geometric figures, intersect means to overlap or cross at more than one point.

When two planes intersect, they meet at a line.

Sets may also intersect, or share common elements. This Venn diagram shows the intersection of multiples of 2 and 3 within the set of whole numbers from 1 to 12.

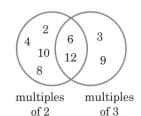

multiples of 2   multiples of 3

See also parallel lines, parallel planes.

intersecting lines

Intersecting lines are lines that cross at one point.

See also intersect.

intersecting planes

Intersecting planes are planes that meet at a line.

See also intersect.

inverse operations

Inverse operations are two operations, each of which "undoes" the other. Addition and subtraction are inverse operations.

Also called opposite operations.

See also opposite operations.

irrational number

An irrational number is a number that cannot be written in the form of a common fraction. When an irrational number is written in decimal form, it is written with the three dots (…) to show that it does not end.

Example

3.14159… (a value for π) is an irrational number.

See also nonterminating decimal.

irregular polygon

An irregular polygon has sides and angles that are not congruent.

See also polygon.

isosceles trapezoid

An isosceles trapezoid is a trapezoid with nonparallel sides of equal length.

See also trapezoid.

isosceles triangle

An isosceles triangle is a triangle that has two sides of equal length.

See also triangle.

jiffy

In computer engineering, jiffy is a term used to describe a very short span of time, from $\frac{1}{50}$ second to as little as 0.01 second (10 milliseconds) or even less.

In physics, jiffy is also a unit of time used to label a light-centimeter, or the time needed for light to travel the distance of one centimeter.

In everyday language, "in a jiffy" usually means in a moment, or shortly.

See also light-year.

jump strategy

The jump strategy is a way of adding or subtracting numbers by hopping by increments of 10s or 1s on a number line.

Examples

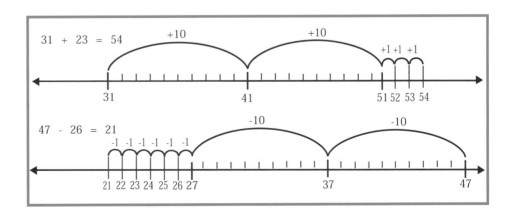

key

The key to a graph or a map is the part that explains the symbols.

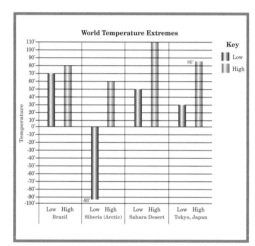

Also called legend.

kilo-

Kilo- is a prefix meaning thousand.

Examples

1 kilometer = 1,000 meters
1 kiloliter = 1,000 liters
1 kilogram = 1,000 grams

See also metric system of measurement.

kilogram (kg)

A kilogram is a unit of weight in the metric system of measurement.

1 kilogram = 1,000 grams

kiloliter (kL)

A kiloliter is a unit of capacity in the metric system of measurement.

$$1 \text{ kiloliter} = 1,000 \text{ liters}$$

Example

A cube that measures 1 meter on each edge has a capacity of 1 kiloliter. A kiloliter of water weighs a metric ton.

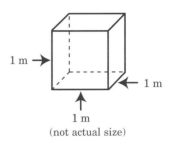

1 m
1 m
1 m
(not actual size)

kilometer (km)

A kilometer is a unit of length in the metric system of measurement. It is about 0.6 of a mile.

$$1 \text{ kilometer} = 1,000 \text{ meters}$$

kite

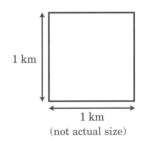

A kite is a quadrilateral with two pairs of sides that are of equal length. These equal sides share a vertex, or "corner." No two sides are parallel.

See also quadrilateral.

km²

Read as *square kilometer.*

A km² is equal to the area enclosed by a square that measures 1 kilometer by 1 kilometer.

Large land and water areas are measured in km².

1 km
1 km
(not actual size)

?? Did You Know ??

The area of 25 city blocks is about 1 km².

The Pacific Ocean, the largest ocean on Earth, has a surface area of about 165,250,000 km².

See also square unit.

latitude

Latitude is a distance north or south of the equator. This distance is measured in degrees (°).

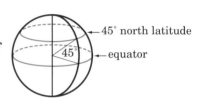

?? Did You Know ??

The equator is 0° latitude, the North Pole is 90° north latitude, and the South Pole is 90° south latitude.

Both Minneapolis, Minnesota (USA), and Bordeaux, France, are located about 45° north latitude. This latitude is halfway between the equator and the North Pole.

See also degrees, longitude.

LCD
(least common denominator)

The least common denominator (LCD) for two or more common fractions is the least common multiple of the denominators.

There are several ways to find the least common denominator of two or more fractions. One way is by listing multiples.

Examples
What is the least common denominator of $\frac{1}{4}$ and $\frac{1}{6}$?

  Multiples of 4:   4, 8, 12, …
  Multiples of 6:   6, 12, …

The least common denominator of $\frac{1}{4}$ and $\frac{1}{6}$ is 12.

LCD
(least common denominator)

(continued)

What is the LCD of $\frac{3}{4}$, $\frac{2}{5}$, and $\frac{1}{2}$?

      Multiples of 4:   4, 8, 12, 16, 20, ...
      Multiples of 5:   5, 10, 15, 20, ...
      Multiples of 2:   2, 4, 6, 8, 10, 12, 14, 16, 18, 20, ...

The LCD of $\frac{3}{4}$, $\frac{2}{5}$, and $\frac{1}{2}$ is 20.

This is written as LCD $(\frac{3}{4}, \frac{2}{5}, \frac{1}{2}) = 20$

See also common multiple, LCM (least common multiple).

LCM
(least common multiple)

The least common multiple (LCM) of two or more numbers is the smallest whole number (except zero) that is a multiple of each number. There are several ways to find the least common multiple of two or more numbers.

Listing multiples is one way to find the LCM.

Example

To find the LCM of 2, 6, and 9, list the nonzero multiples of each number until you find one that is common to all three numbers.

      2, 4, 6, 8, 10, 12, 14, 16, 18, ...
      6, 12, 18, ...
      9, 18, ...

The LCM of 2, 6, and 9 is 18.

This is written as LCM (2, 6, 9) = 18.

Prime factorization is another way to find the LCM.

Example

To find the LCM of 4 and 6, first write the prime factorization of each number.

      $4 = 2 \times 2$
      $6 = 2 \times 3$

Then multiply the prime factors of both numbers together. Include each prime factor as many times as it appears in 4 or 6. This Venn diagram will help you see how many times to include each factor.

      $2 \times 2 \times 3 = 12$

LCM (4, 6) = 12

See also common multiple, LCD (least common denominator), multiple.

length

Length usually refers to the measure of the distance from one end of any object or space to the other. For a rectangle, length is considered the longer of the two dimensions.

?? Did You Know ??

A boa constrictor can grow to a length of 10 to 15 feet.

less than (<)

Read as *is less than*.

The symbol for less than (<) is used to compare two unequal numbers when the lesser number is written first.

Examples

$7 < 10$ — Read as *seven is less than ten*.

$\frac{1}{3} < \frac{1}{2}$ — Read as *one third is less than one half*.

$^-5 < ^-3$ — Read as *negative five is less than negative three*.

See also equality, greater than (>), inequality.

less than or equal to (≤)

Read as *is less than or equal to*.

The symbol for less than or equal to (≤) is used to compare two numbers when the first number is less than or equal to the second number.

Examples

$5 \leq 10$ — Read as *five is less than or equal to ten*.

$6 \leq 6$ — Read as *six is less than or equal to six*.

See also equality, inequality.

light-year

A light-year is the distance light travels through space in one year. A light-year is almost six trillion (6,000,000,000,000) miles.

like fractions

Like fractions are fractions with the same denominator.

Example

$\frac{5}{12}$ and $\frac{1}{12}$ are like fractions.

Also called similar fractions.
See also unlike fractions.

line

A line is a collection of points along a straight path extending infinitely in both directions.

See also line segment, ray.

line graph

A line graph is a graph using line segments to connect points.
Line graphs usually show changes that happen over a period of time.

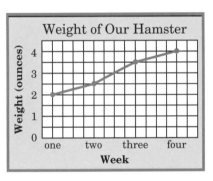

double line graph
A double line graph is a set of two line graphs shown on the same grid. Double line graphs help in comparing two sets of information.

The double line graph pictured here compares temperature extremes in different parts of the world.

See also double line graph.

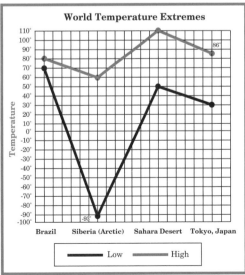

line of symmetry

A line of symmetry is a line that divides a figure into two congruent parts (fig. 1). These parts are mirror images.

Some figures have more than one line of symmetry. In figure 2, all the lines of symmetry for the equilateral triangle and square are shown. For the circle, only four of an infinite number of lines of symmetry are shown.

See also symmetry.

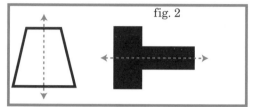

fig. 1

fig. 2

line plot

A line plot is a graph on a number line that shows each item of information. It can serve as a rough draft of a graph. It has the shape of a bar graph.

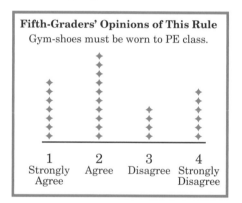

Fifth-Graders' Opinions of This Rule
Gym-shoes must be worn to PE class.

| 1 Strongly Agree | 2 Agree | 3 Disagree | 4 Strongly Disagree |

line segment

A line segment is a part of a line having two endpoints.

See also line, ray.

$\overline{AB}$

A   B

Read as *line segment AB*.

line symmetry

A plane figure has line symmetry if it can be divided into two congruent parts that reflect each other. In other words, the two parts are mirror images. Some figures have many lines of symmetry.

Also called reflectional symmetry.
See also line of symmetry, rotational symmetry, symmetry.
Related word symmetrical.

liquid measure

Liquid measure is a measure of capacity, or the amount of liquid a container will hold. Customary units used for liquid measure include cups, pints, quarts, and gallons. Metric units used for liquid measure include milliliter and liter.

See also capacity, dry measure.

liter (L)

Liter is the basic unit of capacity in the metric system of measurement.

1 liter = 1,000 milliliters

?? Did You Know ??

A liter of water weighs 1 kilogram.
Soda can be bought in 2-liter bottles.
A 2-liter bottle of soda weighs about
2 kilograms.

See also capacity, metric system of measurement.

longitude

Longitude is a distance east or west of an imaginary line on Earth's surface. This distance is measured in degrees (°).

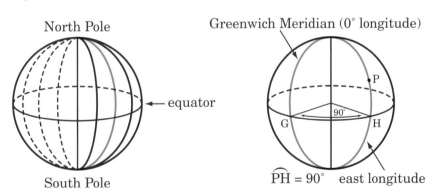

Astronomical Observatory in Greenwich, England, was chosen to be 0° longitude. The International Date Line is halfway around the world, at 180° longitude. The Earth is divided into 24 time zones, or one time zone for each hour of the day. Each time zone is about 15° of longitude, and the International Date Line marks the place that a new day begins.

See also latitude, time zone.

©National Maritime Museum, London, #D6854

loss

In business transactions, loss is the difference in expenses and income when expenses are greater than income.

Example

Pedro made some crafts to sell at the craft fair.
He paid $15.43 for supplies.
He sold all the crafts for only $12.00.
Not counting the time he spent making the crafts or the additional supplies he borrowed from his mother, what was Pedro's loss?

Pedro's expenses were $15.43.
Pedro's income was $12.00.
15.43 − 12.00 = 3.43
Pedro's loss was $3.43.

See also profit.

lowest terms

A common fraction is in lowest terms if the numerator and denominator have no common factor other than 1.

Also called lowest terms of a fraction, simplest form.
See also simplest form.

m²

Read as *square meter*.

A m² is equal to the amount of area enclosed by a square that measures 1 meter by 1 meter.

Also written as square meter.
See also square unit.

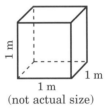

1 m

1 m

(not actual size)

m³

Read as *cubic meter*.

A m³ is equal to the volume of a cube that measures 1 meter (100 cm or 10 dm) on each edge.

Also written as cubic meter.
See also cubic unit.

1 m

1 m

1 m

(not actual size)

mapping

A mapping is a function, or a relation between two sets in which each member of the first set is paired with one and only one member of the second set.

Mapping is also a way to show a function.

See also function, function rule.

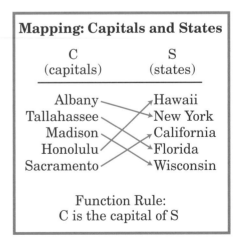

Mapping: Capitals and States

C (capitals)	S (states)
Albany	Hawaii
Tallahassee	New York
Madison	California
Honolulu	Florida
Sacramento	Wisconsin

Function Rule:
C is the capital of S

map scale

A map scale is a ratio between the dimensions on a map and the dimensions of the area represented by the map.

See also scale.

mass

Mass is the amount of matter in an object. It is usually measured in grams and kilograms. The mass of an object remains the same regardless of location.

This astronaut has the same mass, whether on Earth or in space.

See also weight.

mean

The arithmetic mean of a set of numbers is the most common form of average. To find the mean, first find the sum of the set of numbers and then divide by the number of numbers in the set.

Example

Rosa went bowling last evening. She bowled three games. Her scores were:

Game 1	130
Game 2	124
Game 3	133

What was the mean (or average) of her scores for the evening?

130 + 124 + 133 = 387
387 ÷ 3 = 129

The mean of Rosa's bowling scores for the evening was 129.

See also average.

measurement

Measurement is the comparison between an attribute of an object such as length, capacity, or weight and a unit of measure for that attribute.

The measurement of the object is also the number resulting from the comparison.

measurement
(continued)

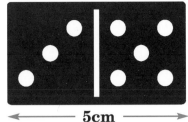

Example

The attribute that is being measured is length. The unit of length is centimeters. This measurement, like all other measurements, is approximate. For every unit of measurement, there is always a smaller, more precise unit that could be used.

5cm

There are many other kinds of measurement. Some of these are area, time, temperature, and angle measurement.

See also attribute, customary system of measurement, metric system of measurement.

median

The median is the middle number for a set of data when the data are arranged in order from least to greatest or greatest to least. To find the median of a set of numbers, arrange the numbers in order and then find the middle number.

Example when there is a middle number

> Set of numbers (arranged in order): 1, 3, 6, 9, 12
> Median: 6

Example when there is no single middle number

If there is no single middle number, the median is the arithmetic mean of the two middle numbers.

> Set of numbers: 1, 2, 4, 5, 7, 8
> $(4 + 5) \div 2 = 4.5$
> Median: 4.5

See also average.

mental computation

Mental computation is doing addition, subtraction, multiplication, or division "in one's head" without using paper and pencil or a calculator. There are many ways to do mental computation. A few of them include using basic facts, front-end estimation, and compatible numbers.

Example using basic facts

A 6-pack of candy bars costs $2.40.
How much does each candy bar cost?
> $2.40 \div 6 = ?$

One solution strategy is to think $24 \div 6 = 4$ (using a basic fact)
So $2.40 \div 6 = .40$
Each candy bar costs $0.40.

mental computation
(continued)

Example front-end estimation

Bob and Joan traveled 313 miles on the first day
of their trip and 178 miles on the second day.
How many miles did they travel in both days?

$$313 + 178 = ?$$

One solution strategy is to think

$$300 + 100 = 400$$
$$10 + 70 = 80$$
$$3 + 8 = 11$$
$$400 + 80 + 11 = 491$$

Bob and Joan traveled 491 miles in both days.

Example compatible numbers

The 32 students in Mr. Clark's class surveyed their favorite fast
foods. Of these students, 13 listed pizza as their favorite.
How many listed some other food?

$$32 - 13 = ?$$

One solution strategy is to think 32 is close to 33.
33 and 13 are compatible numbers.
That is, 33 − 13 is easy to compute mentally.

$$33 - 13 = 20$$

33 is 1 more than 32, so I will need to subtract 1 from 20.

$$20 - 1 = 19$$

In Mr. Clark's class, 19 students listed some other fast food
besides pizza as a favorite.

Also called mental arithmetic, mental math, and mental
mathematics.

See also compatible numbers, estimation strategies,
front-end estimation.

meter (m)

Meter is a unit of length in the metric system of measurement.

1 meter = 100 centimeters.

See also metric system of measurement, length.

metric system of measurement

The metric system of measurement is a base-ten system
of measurement. It was developed in France in the late
1700s and is the major system of measurement in most
of the world. One exception is the United States, where the
customary system of measurement is used in most everyday
situations.

Some of the units of the metric system are shown in the
following table.

	Length	Weight	Capacity
Group 1	millimeter centimeter decimeter	milligram centigram decigram	milliliter centiliter deciliter
Base Unit	meter	gram	liter
Group 2	dekameter hectometer kilometer	dekagram hectogram kilogram	dekaliter hectoliter kiloliter

base units
The base unit for length is meter.
The base unit for weight is gram.
The base unit for capacity is liter.

group 1 prefixes
Group 1 uses the Latin prefixes *milli-*, *centi-*, and *deci-*.
These prefixes added to words expressing measurement show
that the base unit has been divided by some multiple of 10.

1 millimeter = 0.001 meter	1 milliliter = 0.001 liter
1 centimeter = 0.01 meter	1 centiliter = 0.01 liter
1 decimeter = 0.1 meter	1 deciliter = 0.1 liter

1 milligram = 0.001 gram   All the units in Group 1 are
1 centigram = 0.01 gram   smaller than the base unit.
1 decigram = 0.1 gram

group 2 prefixes
Group 2 uses the Greek prefixes *deka-*, *hecto-*, and *kilo-*.
These prefixes show that the base unit has been multiplied
by some multiple of 10.

1 dekameter = 10 meters	1 dekaliter = 10 liters
1 hectometer= 100 meters	1 hectoliter= 100 liters
1 kilometer = 1,000 meters	1 kiloliter = 1,000 liters

1 dekagram = 10 grams   All the units in Group 2 are
1 hectogram = 100 grams   larger than the base unit.
1 kilogram = 1,000 grams

metric system of measurement
(continued)

Also called metric measurement system and metric system.

See also capacity, Celsius (°C) temperature scale, centigram, centiliter, centimeter, cm², cubic unit, decigram, deciliter, decimeter, dekagram, dekaliter, dekameter, gram, hectogram, hectoliter, hectometer, kilogram, kiloliter, kilometer, km², liter, m², meter, metric ton, milligram, milliliter, millimeter.

metric ton (t)

Metric ton is a unit of weight in the metric system of measurement.

> 1 metric ton = 1,000 kilograms

The blue whale is the world's largest animal. It is the length of 3 school buses and may weigh as much as 135 metric tons. This weight is more than 25 times as much as an elephant, the largest land animal and the second-largest animal in the world.

See also metric system of measurement, ton.

mi²

Read as *square mile*.

A mi² is equal to the area enclosed by a square that measures 1 mile by 1 mile.

See also square unit.

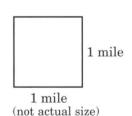

1 mile
1 mile
(not actual size)

mile (mi)

A mile is a unit of length in the customary system of measurement.

> 1 mile = 5,280 feet or 1,760 yards

See also customary system of measurement.

millennium

plural millenniums or millennia

Millennium is a measure of time.

1 millennium = 1,000 years

?? Did You Know ??

The bristlecone pine tree, found in Utah and only five other western states, may be the Earth's oldest living tree. Several bristlecone pine trees are believed to be more than 4 millenniums old!

milli-

Milli- is a prefix meaning one thousandth.

1 millimeter = 0.001 meter
1 milliliter = 0.001 liter
1 milligram = 0.001 gram

milligram (mg)

A milligram is a unit of weight in the metric system of measurement.

1,000 milligrams = 1 gram

See also metric system of measurement, weight.

milliliter (mL)

A milliliter is a unit of capacity in the metric system of measurement.

1,000 milliliters = 1 liter

See also capacity, metric system of measurement.

millimeter (mm)

A millimeter is a unit of length in the metric system of measurement.

1,000 millimeters = 1 meter

See also length, metric system of measurement.

million

A million is equal to 1,000 thousands. In standard form, one million is written as 1,000,000. With an exponent, one million may be written as
$$1 \times 10^6 \text{ (or simply as } 10^6).$$

See also decimal numeration system.
Related word millionth.

minute (min)

A minute is a unit for measuring short lengths of time.
$$60 \text{ seconds} = 1 \text{ minute} \qquad 60 \text{ minutes} = 1 \text{ hour}$$

A minute is also a unit for measuring angles.
$$60 \text{ minutes} = 1 \text{ degree of angle measure}$$

See also second.

missing addend

A missing addend is the number that must be added to one given number to equal another given number.

Example

Leila saved \$60. She wants a new bicycle that costs \$100. How much more money does she need to buy the bicycle?
$$60 + \square = 100$$
The missing addend is 40. Leila needs \$40 to buy the bicycle.

mixed number

A mixed number has both a whole number and a fractional part. The fractional part may be a common fraction, or it may be in decimal form.

Examples

$$1\tfrac{1}{2} \qquad 3\tfrac{3}{5} \qquad 23\tfrac{4}{7} \qquad 2.5 \qquad 4.03$$

See also decimal mixed number.

mode

The mode is the number that occurs most often in a set of numbers. Some sets of numbers have more than one mode, and some sets have no mode.

mode
(continued)

Example one mode for a set of numbers

Set of numbers: 1, 3, 3, 4, 7, 8
Mode: 3

Example more than one mode for a set of numbers

Set of numbers: 1, 3, 3, 4, 7, 7, 8
Modes: 3, 7

Example no mode for a set of numbers

Set of numbers: 1, 3, 4, 7, 8
Mode: none

See also average.

motif

A motif is a shape or pattern that is repeated in a design. Can you find the motif, or repeated pattern, in this quilt?

See also tessellation.

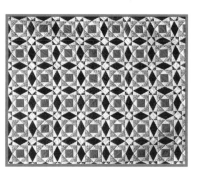

multiple

A multiple of a whole number is the product of that number and another whole number.

Examples

Multiples of 4 include 0, 4, 8, 12,. . . .
   (0 × 4 = 0; 1 × 4 = 4; 2 × 4 = 8; 3 × 4 = 12;. . . .)
Multiples of 6 include 0, 6, 12, 18,. . . .
   (0 × 6 = 0; 1 × 6 = 6; 2 × 6 = 12; 3 × 6 = 18;. . . .)

See also common multiple, least common multiple.
Related words multiplication, multiply.

multiplication

Multiplication is one of the four basic operations on numbers. (The other basic operations are addition, subtraction, and division.)

The numbers that are multiplied are called factors, and the answer is called the product.

multiplication
(continued)

Example

In $4 \times 9 = 36$,
  4 and 9 are factors.
  36 is the product.

Multiplication can be explained by using several different situations. Some of these situations are repeated addition, array, area, and combinations.

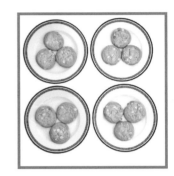

Example repeated addition

There are 3 cookies on each of 4 plates.
  $4 \times 3 = 3 + 3 + 3 + 3 = 12$
There are 12 cookies in all.

Example array

There are 2 rows and 7 columns.
  $2 \times 7 = 14$
There are 14 ice cubes in this tray.

Example area

  $3 \times 4 = 12$
The area of the rectangle is 12 square units.

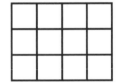

Example combinations

Danae was going on a trip. She wanted to pack as many outfits as she could, using as few items of clothing as possible. She packed four shirts: turtleneck, knit top, T-shirt, and sweatshirt. She packed three pairs of pants: jeans, sweatpants, and slacks. How many different outfits can she make?

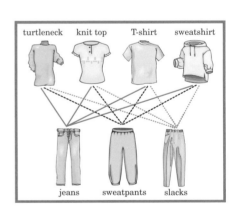

turtleneck knit top T-shirt sweatshirt

jeans sweatpants slacks

  $4 \times 3 = 12$ Danae could make 12 outfits.

A common procedure for doing multiplication is called the partial products algorithm.

See also area, array, basic facts, factor, partial products algorithm, product.

Related words multiple, multiplicative, multiply.

multiplication facts

The multiplication facts are the 100 multiplication combinations of one-digit numbers.

Examples

$8 \times 1 = 8$ $7 \times 5 = 35$

See also basic facts, multiplication table.

multiplication sentence

A multiplication sentence is a number sentence used to express multiplication.

Examples

$6 \times 4 = 24$
6 and 4 are factors.
$\times$ is the symbol for multiplication.
= indicates that the amount on either side is equal, or has the same value.
24 is the product.

See also equal, factor, number sentence, product.

multiplication table

A multiplication table is a table that organizes the 100 basic multiplication facts.

See also basic facts, reciprocal.

X	0	1	2	3	4	5	6	7	8	9
0	0	0	0	0	0	0	0	0	0	0
1	0	1	2	3	4	5	6	7	8	9
2	0	2	4	6	8	10	12	14	16	18
3	0	3	6	9	12	15	18	21	24	27
4	0	4	8	12	16	20	24	28	32	36
5	0	5	10	15	20	25	30	35	40	45
6	0	6	12	18	24	30	36	42	48	54
7	0	7	14	21	28	35	42	49	56	63
8	0	8	16	24	32	40	48	56	64	72
9	0	9	18	27	36	45	54	63	72	81

$3 \times 7 = 21$ $7 \times 3 = 21$

mutually exclusive events

Mutually exclusive events are events that cannot occur at the same time.

Examples

Being 10 years old and 14 years old at the same time.
Being in Chicago and Dallas at the same time.
Rolling an odd number and an even number on one die at the same time.

natural numbers

Natural numbers are the numbers 1, 2, 3, 4, 5, …
They go on without end. (Notice that 0 and negative numbers are not included in the set of natural numbers.)

Also called counting numbers.

negative integer

A negative integer is an integer that is less than zero.

See also integers.

negative number

A negative number is any number less than zero.
Negative numbers are written with a negative sign (⁻).

Examples

The temperature outside was ⁻4°C.
The lowest point on Earth is found in the Pacific Ocean near Guam. Its elevation has been measured at ⁻35,840 feet, or 35,840 feet below sea level.
The change in interest rates on home loans was ⁻0.1%.

net

A net is a pattern that can be cut and folded to make a space figure. The Earth icosahedron on page 200 of this book is an example of a net.

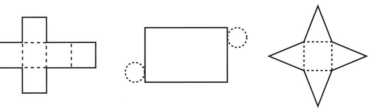

What space figures can be made from these nets?

nonterminating decimal

A nonterminating decimal goes on without end.

Examples

$\frac{1}{3}$ may be written as the decimal fraction $0.\overline{3}$.
The bar indicates that the 3 continues to repeat itself indefinitely.
$0.\overline{3}$ or 0.333... is a nonterminating decimal.

$\frac{3}{11}$ may be written in decimal form as $0.\overline{27}$.
In this number, the 27 repeats.
$0.\overline{27}$ or 0.2727... is a nonterminating decimal.

See also terminating decimal.

not equal to (≠)

Read as *is not equal to*.

The sign for not equal to (≠) is used to indicate that two numbers do not have the same value.

Example

$4 \neq 5$ Read as *four is not equal to five*.

See also inequality.

number

A number can be thought of as a concept or an idea that indicates how many or how much.

The number of each set of objects pictured below can be named by the numeral 3.

There are 3 candy apples, 3 muffins, and 3 people reading.

A number may be named in more than one way:
with words, in standard form, and in expanded form.

Example

Word form: two hundred thirty-four
Standard form: 234
Expanded form: 200 + 30 + 4

?? Did You Know ??

The concept of number is so abstract that a general
mathematical definition has not been written for this
term. Mathematicians worked for many, many years to
find the words to define *number* accurately. Finally, they
decided that the best way was to use a specific example.
Although many people use the word *number* to mean
numeral, they are not the same. Numerals are used to
represent numbers, and one number can be represented
with many different numerals. For example, some of the
ways to represent three are 3, III (Roman numerals),
and 2 + 1.

See also approximate number, cardinal number, compatible
numbers, composite number, counting numbers, decimal mixed
number, even number, expanded form, figurate number,
fractional number, integers, irrational number, mixed number,
numeral, odd number, ordinal number, pentagonal number,
perfect number, polygonal numbers, prime number, rational
number, rectangular number, square number, standard form,
triangular number, whole numbers, word form.

number line   A number line is a line (or line segment or ray) on which
numbers are assigned points.

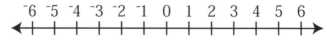

$^-6$ $^-5$ $^-4$ $^-3$ $^-2$ $^-1$ 0 1 2 3 4 5 6

**number
pattern**

A number pattern is a series of numbers arranged or repeated
in some order or design.

Examples

Pattern of odd numbers: 1, 3, 5, 7, 9, 11, ...
Pattern of multiples of 3:   0, 3, 6, 9, 12, 15, ...
Pattern of powers of 2:   2, 4, 8, 16, 32, 64, ...
Fibonacci sequence: 1, 1, 2, 3, 5, 8, 13, ...

**number
pattern**
(continued)

?? Did You Know ??

The Fibonacci sequence is named for an Italian mathematician named Leonardo Fibonacci, who lived about 800 years ago. This sequence can be found in many living things, such as this pine cone. If you count the seeds in the spirals of the pine cone, you will find adjacent numbers in the Fibonacci sequence.

Also called number sequence.

See also pattern.

**number
sense**

Number sense is an understanding of numbers. It includes ideas related to number meanings, number relationships and size, and the relative effects of operations on numbers.

?? Did You Know ??

Developing number sense is a lifelong process and is never completely learned. For example, many adults have difficulty giving real-life examples for numbers such as 100, 1,000, and so on.

**number
sentence**

A number sentence is used to express an arithmetic operation. It is written by using numerals to represent numbers.

Examples with whole numbers

Addition sentence:	$5 + 3 = 8$
Subtraction sentence:	$6 - 4 = 2$
Multiplication sentence:	$7 \times 8 = 56$
Division sentence:	$8 \div 2 = 4$

Number sentences can also be written with fractions, decimals, negative numbers, etc.

See also addition sentence, division sentence, equation, multiplication sentence, subtraction sentence.

numeral

A numeral is a symbol that names a number. There are many different ways numerals can be used to name the same number.

Examples *(for 8)*

With a digit:	8
With the addition operation:	5 + 3
With the division operation:	16 ÷ 2
With Roman numerals:	VIII
With tally marks:	𝗧𝗛𝗟 III

See also Roman numerals.

numeration system

A numeration system is an organized way of writing the numerals for numbers.

?? Did You Know ??

The Egyptians developed a system based on groups of ten about 5,000 years ago. The decimal numeration system we now use, also called the Hindu-Arabic numeration system, combines groups of ten, place value, and the use of zero to allow for many complex mathematical ideas and operations.

Egyptian Numeration System

1 = | (staff)   1,000 = 🪷 (lotus flower)

10 = ∩ (heel)   10,000 = 𝑓 (bent finger)

100 = 𝟡 (coil)   100,000 = ⌒ (tadpole)

1,000,000 = 𝕐 (man with raised arms

|||| ∩∩∩ / ∩∩   𝟡 / 𝟡

4   +   50   +   200

---

numeration system
(continued)

See also decimal numeration system, Hindu-Arabic numeration system, Roman numeration system.

numerator

The numerator is the name for one of the two terms of a common fraction. (The other term is called the denominator.) The numerator is the "counting" number. It tells how many fractional parts in a whole or set have been counted.

$$\frac{3}{4}$$ ← numerator
← fraction bar
← denominator

Numerator also has other meanings. For example, it is the first term of a ratio expressed as a common fraction.

Notice that the numerator is written above the fraction bar.

Examples

These photos of half a pie and a bagel cut in half illustrate the fraction $\frac{1}{2}$. 1 is the numerator.

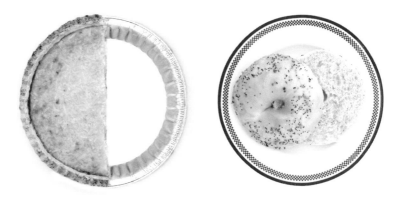

See also common fraction, denominator, fraction bar, ratio, terms.

oblique

Oblique may be used to describe angles. It may also describe space figures or three-dimensional figures.

Examples

Oblique angles are angles that are not right angles. They are either acute or obtuse.

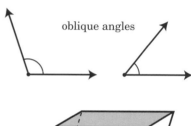

oblique angles

Oblique space figures
This rectangular prism is oblique. The bases are not directly in line with each other.

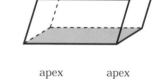

Both the cone and the pyramid are oblique. The apex is not directly in line with the center of the base.

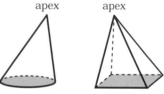

apex   apex

obtuse angle

An obtuse angle is an angle that measures more than 90° but less than 180°.

?? Did You Know ??

Obtuse comes from a Latin word meaning blunted or dull (the opposite of acute or sharp). The vertex of an obtuse angle is "dull" when compared with the vertex of an acute angle.

See also acute angle, angle, right angle, straight angle.

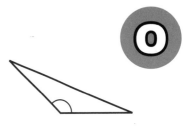

obtuse triangle

An obtuse triangle is a triangle with an obtuse angle (an angle greater than 90°).

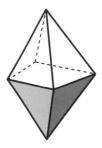

See also acute triangle, right triangle, triangle.

octagon

An octagon is a polygon with eight sides. (*Octa-* is a Greek prefix meaning eight.)

Stop signs are in the shape of an octagon.

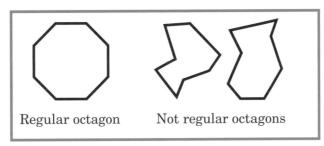

Regular octagon   Not regular octagons

See also polygon.

octahedron

An octahedron is a space figure with eight faces. If each face is an equilateral triangle, then it is one of the five regular polyhedra.

See also polyhedron.

odd number

An odd number is a whole number that cannot be grouped in twos. There is always a remainder of 1 when an odd number is divided by 2.

Odd numbers have 1, 3, 5, 7, or 9 in the ones place.

See also even number.

one-dimensional

A line, ray, or line segment is one-dimensional.
The one dimension of a line, ray, or line segment is length.

Line   Ray   Line segment

See also geometric figure, three-dimensional, two-dimensional, zero-dimensional.

Related word dimension.

open sentence

An open sentence is a mathematical sentence for which more information is needed. The information is needed to be able to tell whether the sentence is true or false.

Example

$$4 + ? = 9$$

statement

If a number is used to replace the ? in the example, the open sentence becomes a statement that can be identified as true or false. Writing an open sentence can be helpful in problem solving.

Examples

Michelle spent $15 at the camp store.
She bought a T-shirt for $8.
How much did she spend for other items?

$$8 + ? = 15$$ She spent $7 for other items.

After Michelle's trip to the camp store, she had $0.75 left in coins. She had only dimes and nickels, and she had the same number of each.
How many dimes and how many nickels did she have?

Since a dime is equal to $.10, and a nickel is equal to $.05, we can write the following open sentence.

The ? represents the number we do not know.
$$(? \times 10) + (? \times 5) = 75$$

By trying different numbers for both ?, we can solve the problem, making the open sentence a true statement.

operation

An operation is an action upon numbers that results in a single number. The arithmetic operations of addition, subtraction, multiplication, and division are operations on two numbers.

Examples

$$4 + 2 = 6 \qquad 7 \times {}^-3 = {}^-21$$
$$12 - 3 = 9 \qquad \frac{1}{4} \div \frac{1}{2} = \frac{1}{2}$$

An operation can also be performed on a single number.

Example

$$4^2 = 16$$

See also addition, basic operations, division, exponent, multiplication, subtraction.

opposite integers

Opposite integers are two integers that are the same distance from 0 on the number line, but in the opposite direction.

Example

⁻4 and ⁺4 are opposite integers.

The set of integers consists of all the whole numbers, their opposites, and zero.

See also integers, opposite of a number.

opposite of a number

The opposite of a number is a number on the number line that is the same distance from 0 as the given number, but in the opposite direction. The sum of a number and its opposite is 0.

Examples

⁺5 is the opposite of ⁻5, and ⁻5 is the opposite of ⁺5. (⁺5 and ⁻5 are called opposites, or opposite integers.)

⁺3.5 and ⁻3.5 are opposites, but they are not opposite integers.

Also called additive inverse.

See also additive inverse.

opposite operations

Opposite operations are two operations, each of which "undoes" the other. Addition and subtraction are opposite operations.

Example

$$4 + 3 = 7$$
$$7 - 3 = 4$$

Multiplication and division are also opposite operations.

Example

$$6 \times 5 = 30$$
$$30 \div 5 = 6$$

Also called inverse operations.

See also fact family.

opposite sides and angles of a quadrilateral

For a quadrilateral, opposite sides do not have a common vertex.

Line segments AB and CD are opposite sides of this quadrilateral. Line segments AD and BC are also opposite sides.

For a quadrilateral, opposite angles do not have a common side.

Angle A and angle C are opposite angles in this quadrilateral. Angle B and angle D are also opposite angles.

opposite sides and angles of a triangle

For a triangle, a side and an angle are opposite each other if the other two sides are sides of the angle.

Line segment AB is opposite angle C.

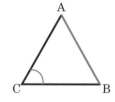

ordered pair

An ordered pair is a set of two numbers that identifies a location on a map or in a coordinate system.

Sometimes a map will use a letter to represent one of the numbers in the ordered pair.

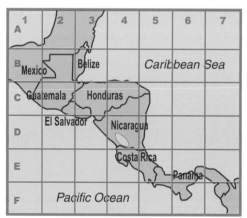

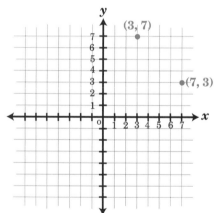

On this map of Central America, some of the coordinates locate the countries of Mexico (B, 1), Belize (B, 3), Guatemala (C, 2), Honduras (C, 3), and Nicaragua (D 4).

On the coordinate grid, locate (3, 7). Now locate (7, 3).

Notice how the order of the numbers in the ordered pair makes a difference in the location of the point on the coordinate grid.

See also coordinate system.

order of operations

Order of operations is an order, agreed upon by mathematicians, for performing operations to simplify expressions. This order is as follows:

First, perform all operations within parentheses or other grouping symbols.
Second, simplify expressions involving exponents.
Third, multiply and divide in order from left to right.
Fourth, add and subtract in order from left to right.

Examples

$$10^2 \div (8 \times 2 - 6) + 1 \qquad 3 \div 3 + 3 \times 3 - 3$$
$$10^2 \div (16 - 6) + 1 \qquad 1 + 9 - 3$$
$$10^2 \div 10 + 1 \qquad 10 - 3$$
$$100 \div 10 + 1 \qquad 7$$
$$10 + 1$$
$$11$$

Some people use the mnemonic "Please Excuse My Dear Aunt Sally" to remember this order: Parentheses, Exponents, Multiplication and Division, Addition and Subtraction.

order property

The order property means that changing the order of the numbers used in an operation does not change the result of that operation. Addition and multiplication have the order property, but subtraction and division do not.

Also called commutative property.

See also commutative property of addition, commutative property of multiplication.

order property of addition

The order property of addition means that changing the order in which numbers are added does not change the sum.

Also called commutative property of addition.

order property of multiplication

The order property of multiplication means that changing the order in which numbers are multiplied does not change the product.

Also called commutative property of multiplication.

ordinal number

An ordinal number is a number used to tell order, or position. Each ordinal number can be paired with a cardinal number. (A *cardinal number* is a whole number that tells how many are in a group.)

Examples

Ordinal Number	Cardinal Number
first	one, or 1
second	two, or 2
third	three, or 3
fourth	four, or 4
tenth	ten, or 10
twenty-first	twenty-one, or 21
hundredth	one hundred, or 100

See also cardinal number.

Related word order.

origin

In a coordinate system, the origin is the point of intersection of the x- and y-axes, which are number lines that intersect at zero (0).

On a number line, the origin is the point assigned to zero.

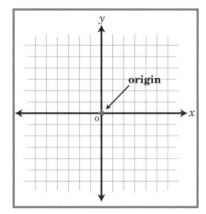

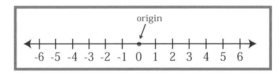

See also cardinal number, coordinate system, number line, x- and y-axes.

ounce (oz)

An ounce is a unit of weight in the customary system of measurement.

16 ounces = 1 pound

fluid ounce

An ounce (sometimes called *fluid ounce*) is also a unit of capacity in the customary system of measurement.

8 ounces (or 8 fluid ounces) = 1 cup

ounce (oz)
(continued)

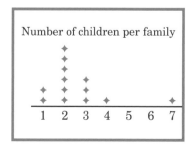

?? *Did You Know* ??

A hen's egg, such as one you might eat for breakfast, weighs about 2 ounces. A hummingbird's egg, however, is very tiny. It weighs about $\frac{1}{50}$ ounce.

See also customary system of measurement, fluid ounce.

outcome

An outcome is a result of a probability experiment.

possible outcomes
There are two possible outcomes of tossing a penny: heads or tails.

equally likely outcomes
If a penny is tossed, heads and tails are equally likely outcomes. That is, either of these two possible outcomes is equally possible. Since there are two equally likely outcomes and heads is one of them, the expected probability of the penny landing heads up is $\frac{1}{2}$. Toss a penny 30 times to see how many times it lands heads up.

favorable outcomes
Let's say the penny landed heads up 18 times. Since you were experimenting to see how many times it would land heads up, there were 18 favorable outcomes for your experiment. The experimental probability was $\frac{18}{30}$, or $\frac{3}{5}$.

See also equally likely outcomes, favorable outcome, possible outcomes, probability.

outlier

On a line plot, an outlier is a data category some distance away from other data categories.

line plot
A line plot is a graph that shows each item of information on a number line.

In this line plot, 7 is an outlier.

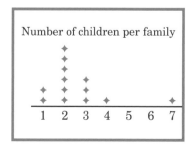

Number of children per family

See also line plot.

oval

An oval is any plane shape that resembles the outline of an egg.

ellipse

An ellipse is a special kind of oval
that is mathematically regular, with two lines of symmetry.

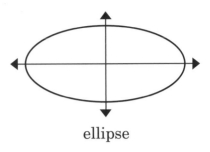

ellipse

?? Did You Know ??

The word *oval* comes from the Latin word *ovum*, which means egg.

The office of the President of the United States is called the Oval Office because of its shape.

See also ellipse.

palindrome

A palindrome is a number that reads the same from left to right and from right to left.

Also called palindromic number.

Examples

878   342.243   2002

?? Did You Know ??

Words and sentences can be palindromes, too. Both the first and last names of this person are palindromes: Bob Otto.

parallel lines

Parallel lines are lines in the same plane that never intersect. Line segments and rays that are parts of parallel lines are also parallel.

These glass building panels are an example of parallel line segments in architecture.

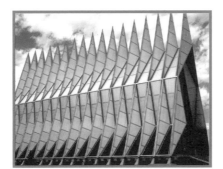

See also intersecting lines, perpendicular lines.
Related word parallelogram.

parallelogram

A parallelogram is a quadrilateral with opposite sides parallel.

See also quadrilateral.
Related word parallel.

parallel planes

Parallel planes are planes in space that never intersect.

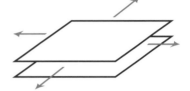

See also plane.

parentheses ( )

Parentheses ( ) are used in mathematics as grouping symbols for operations. When simplifying an expression, the operations within the parentheses are performed first.

Example

$(2 + 3) \times 4$
5×4
20

See also order of operations.

partial products algorithm

The partial products algorithm is a common procedure for multiplying numbers in which one or both factors have more than one digit. A partial product is one part of the result.

Example

$$
\begin{array}{r}
124 \\
\times\ 3 \\
\hline
12 \\
60 \\
\underline{300} \\
372
\end{array}
$$

12   (3×4)
60   (3×20) partial products
300   (3×100)
372   (the sum of the partial products)

See also algorithm, multiplication.

partition division

In partition division, a number is divided into equal parts.

Example

Daynna, Taylor, and John went blueberry picking. Together, they picked 15 pints. When they got ready to go home, they shared the blueberries equally. How many pints did each person have?

$$15 \div 3 = 5$$

Each person's share of blueberries was 5 pints.

Also called partitive division, sharing.

pattern

A pattern is a repeated sequence or design, which may occur everywhere in mathematics as well as in other everyday situations. Patterns in numbers and geometry provide examples of patterns in mathematics.

The picture of repeating arches is an example of a pattern in architecture; the quilt uses a pattern of repeating rhombuses, squares, and triangles.

number patterns in an arithmetic sequence

In some number patterns, the difference between any two consecutive numbers is the same. This kind of pattern is called an arithmetic sequence.

Examples

0, 2, 4, 6, 8, 10, 12, …

This is the pattern of even numbers. Notice that the pattern begins with 0 and the difference in any two consecutive numbers, or terms, is 2.

1, 3, 5, 7, 9, 11, …

This is the pattern of odd numbers. Notice that the pattern begins with 1 and the difference in any two consecutive numbers, or terms, is 2.

pattern
(continued)

0, 6, 12, 18, 24, 30, 36, …
This is the pattern of multiples of 6.
Notice that the pattern begins with 0 and the difference in any two consecutive terms is 6.

Some number patterns are not arithmetic sequences.

Examples

1, 4, 9, 16, 25, …
2, 6, 18, 54, …
1, 1, 2, 3, 5, 8, 13, 21, …

Sometimes arithmetic sequences are called repeating patterns. A pattern such as 1, 4, 9, 16, 25 . . . is sometimes called a growing pattern.

geometry patterns
Some geometry patterns are visual patterns, or patterns you can see.

Example

This picture of a honeycomb is an example of a tessellation. A tessellation is a pattern of shapes that are repeated to fill a plane. The shapes do not overlap and there are no gaps.

Other patterns in geometry are formed by relationships among attributes.

Example

These figures are parallelograms because they fit the pattern formed by these two attributes:
They have four sides.

Opposite sides are parallel.

See also arithmetic sequence, tessellation, tiling.

pentagon

A pentagon is a polygon with five sides.

Example

The Pentagon Building, near Washington, D.C., is one of the world's largest office buildings. From an aerial view, it looks like a pentagon.

See also polygon.

pentagonal number

A pentagonal number is a whole number that can be shown in an array that looks like a pentagon. Arrays for the first five

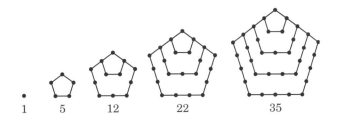

numbers in a pentagonal sequence are shown.

See also polygonal numbers.

percent (%)

Percent (%) is a ratio that represents parts per hundred. It is one way of expressing a fractional number.

Examples

The sales tax in some states is 5 percent. This is an amount is equal to 5 cents in sales tax for every 100 cents (or $.05 per $1.00).

The air we breathe is about 21% oxygen, 78% nitrogen, and 1% argon and other gases.

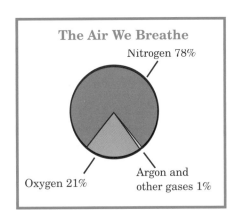

The Air We Breathe

Nitrogen 78%

Oxygen 21%

Argon and other gases 1%

percent (%)
(continued)

Common fractions and decimal fractions are two other ways of expressing fractional numbers.

P

Examples

Common Fraction	Decimal Fraction	Percent
$\frac{1}{10}$	0.1	10%
$\frac{1}{2}$	0.5	50%
$\frac{1}{1}$	1.00	100%
$1\frac{1}{4}$	1.25	125%

See also fractional number, ratio.

Related word percentage.

perfect number

A perfect number is a number for which the sum of its proper factors is equal to the number itself.

proper factor
The proper factors of a number are all its factors except the number itself.

Example

6 is a perfect number because the sum of its proper factors (1, 2, and 3) is equal to 6.

?? Did You Know ??

Perfect numbers have been studied since ancient times. The early Egyptians are believed to have studied perfect numbers. Pythagoras, a famous Greek mathematician and philosopher, studied them, perhaps from a metaphysical (nonphysical) as well as a mathematical perspective. There is still much to learn about perfect numbers, and mathematicians still study them today.

perimeter

Perimeter is the name for the outer edge of a figure or shape. It is also the distance around a shape or figure.

Examples

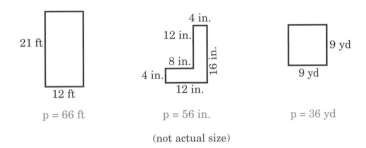

p = 66 ft   p = 56 in.   p = 36 yd

(not actual size)

circumference
Circumference is a special kind of perimeter. It is the distance around a circle.

See also circumference.

period

In large numbers, periods are groups of three digits separated by commas. When large numbers are separated into periods, they are easier to read.

Example

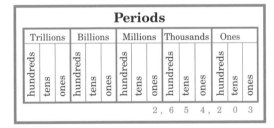

Numbers are sometimes separated into periods by spaces instead of commas.

Example

1 000 000 (Read as *one million*.)

Four-digit numbers may or may not be separated into periods.

Example

2,304 or 2304

perpendicular

Perpendicular means meeting or crossing at right angles. Lines, rays, line segments, and planes can be perpendicular. Perpendicular lines, rays, and line segments are lines or parts of lines that meet or cross at right angles.

Examples

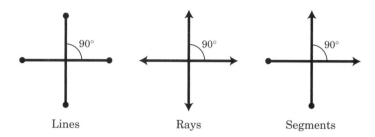

Lines   Rays   Segments

See also perpendicular lines, perpendicular plane.

perpendicular lines

Perpendicular lines are lines that meet at right angles. Sometimes in everyday language, parts of lines (rays and line segments) that meet at right angles are also called perpendicular lines.

See also perpendicular.

perpendicular planes

Perpendicular planes are planes that meet or cross at right angles.

See also perpendicular.

pi (π)

Pi (π) is the ratio of the circumference of a circle to its diameter. The value of pi is 3.1416..., an irrational number. 3.14 and $3\frac{1}{7}$ (or $\frac{22}{7}$) are often used as approximate values for pi. Pi is a constant value. That is, the ratio of the circumference to the diameter is the same for all circles.

This drawing shows the circumference of the circle "stretched out." It is a little more than three diameters in length.

See also circumference, irrational numbers.

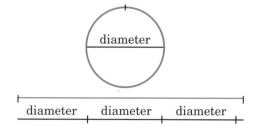

picture graph

A picture graph is a kind of bar graph that uses pictures or drawings to show information.

Automobile Manufacturing: Which States Make the Most Cars	
Michigan	🚗🚗🚗🚗🚗🚗🚗🚗🚗🚗 🚗🚗🚗🚗🚗🚗🚗🚗🚗🚗
Missouri	🚗🚗🚗🚗🚗🚗🚗🚗🚗
Ohio	🚗🚗🚗🚗🚗🚗🚗
Georgia	🚗🚗🚗🚗🚗
Illinois	🚗🚗🚗🚗
Delaware	🚗🚗🚗🚗🚗
New Jersey	🚗🚗

(States)

🚗 stands for 100,000 cars

Also called pictogram, pictograph.
See also bar graph.

pie graph

A pie graph is a graph in the shape of a pie, or circle, that shows how a total amount has been divided into parts. The pie, or circle, represents the total amount. The slices of pie show how the total amount has been divided.

Also called circle graph, pie chart.
See also circle graph.

pint (pt)

Pint is a unit of capacity in the customary system of measurement.

2 cups = 1 pint   2 pints = 1 quart

See also capacity, customary system of measurement.

place value

Place value is the value of a digit determined by its position in a number. In the decimal numeration system, the numeration system we use almost all the time, place value is based on groupings of ten and powers of ten.

Example

In the number 3,487,
8 represents 8 tens and is
said to be in the tens place.

thousands	hundreds	tens	ones
3	4	8	7

place value
(continued)

Computer scientists use a numeration system in which place value is based on groupings of two and powers of two. The first five numbers in this system, called the base two or binary numeration system, are 0, 1, 10, 11, and 100. Their values in base ten are 0, 1, 2, 3, and 4.

Base two has only two digits: 0 and 1. These digits are used to tell the computer what to do. 0 shows that there is no electrical current and 1 shows that the current is on. Through the use of many 0s and 1s, computers can perform many complex operations.

See also decimal numeration system.

plane

A plane is a flat surface that extends in all directions without ending (fig. 1). It is two-dimensional, having length and width but no thickness.

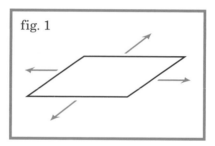

fig. 1

See also plane figure, two-dimensional.

plane figure

A plane figure is a geometric figure that has no thickness (fig. 2). It lies entirely in one plane.

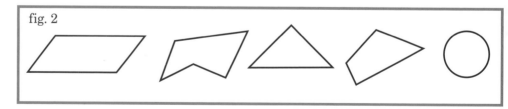

fig. 2

Also called plane shape, two-dimensional figure, two-dimensional shape.

plotting a point

Plotting a point in a coordinate system is locating and marking a point when given its coordinates.

Example

To plot (3, 5), first locate 3 on the *x*-axis and 5 on the *y*-axis. Follow the arrows shown in the drawing.

See also coordinate system, ordered pair.

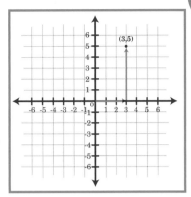

PM

PM is used to label the time between noon and midnight. PM is the abbreviation for *post meridiem*, a Latin phrase meaning "after noon."

?? Did You Know ??

A sundial is an instrument that shows the time of day by the shadow of a pointer on a flat surface. Sundials were used to keep time before mechanical clocks were invented. This photo shows a "human sundial" installed in a schoolyard in Lafayette, Louisiana.

Also written as P.M., PM, pm, and p.m.

See also AM.

point

A point is a location in space.

zero-dimensional
A point is zero-dimensional because it does not occupy space. It has neither length, width, nor height. Because a point has no dimensions, we cannot draw a picture of it. We usually represent a point with a dot.

Mathematicians define geometric figures as sets of points.

See also zero-dimensional.

polygon

A polygon is a closed plane figure formed by line segments.

regular polygon
A regular polygon has all sides of equal length and all angles of equal measure.

irregular polygon
An irregular polygon has sides or angles that are not congruent.

?? Did You Know ??

Polygon comes from Greek words meaning many and angle.

See also plane figure.

polygonal numbers

Polygonal numbers are numbers that can be represented by arrays that look like various plane figures.

Also called figurate numbers.

See also pentagonal number, rectangular number, square number, triangular number.

*plural **polyhedra***

polyhedron

A polyhedron is a closed space figure with faces that are in the shape of polygons.

134

polyhedron
(continued)

regular polyhedra
There are five regular polyhedra.

All the faces of a regular polyhedron are congruent regular polygons, and all the angles are congruent.

Examples

Number of faces	4	6	8	12	20
Name	Tetrahedron	Cube	Octahedron	Dodecahedron	Icosahedron
Sketch					
Net					

See also cube, dodecahedron, icosahedron, octahedron, prism, rectangular prism, tetrahedron, triangular prism.

population

In statistics, a population is the total group from which a sample is taken.

Example

A survey found that vanilla was the favorite ice-cream flavor for a sample of people in one state. The sample was selected from the ice-cream-eating population in that state.

Population also refers to the number of people who live in a specific location.

Examples

The states with the largest and smallest populations are:
    Largest: California, with a population of about 36,000,000
    Smallest: Wyoming, with a population of about 500,000

positive integer

A positive integer is an integer that is greater than 0.

See also integers.

positive number

A positive number is any number greater than 0. (0 is neither positive nor negative.) Positive numbers may be written with the positive sign (⁺). A nonzero number written without a sign is also considered positive.

Examples

Body temperature for humans is 98.6°F.
The elevation of Mt. McKinley is ⁺20,320 feet, or 20,320 feet above sea level.

possible outcomes

Possible outcomes are the possible results of a probability experiment.

Example

When a coin is tossed to see which way it lands, there are two possible outcomes: heads or tails.

See also outcome, sample space.

pound (lb)

A pound is a unit of weight in the customary system of measurement.

Example

1 pound = 16 ounces

The pound is also a unit of money in Great Britain and several other countries.

?? Did You Know ??

The abbreviation *lb* comes from *libra*, a Latin word meaning pound.

See also customary system of measurement, weight.

power of a number

The power of a number is the number of times that number is used as a factor. It is usually indicated by an exponent.

power of a number
(continued)

Examples

$10^2 = 10 \times 10 = 100$, or 10 to the second power.

$2^3 = 2 \times 2 \times 2 = 8$, or 2 to the third power

See also exponent.

prime factor

A prime factor is a prime number that is a factor of a whole number.

Example

The factors of 12 are 1, 2, 3, 4, and 6.
Of these factors, 2 and 3 are prime factors.

See also factor, prime factorization, prime number.

prime factorization

Prime factorization is the expression of a composite number as the product of its prime factors.

Example

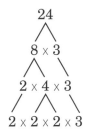 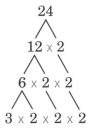 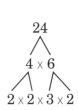

The order of factors varies, but these three factor trees all give the same result. The prime factorization of 24 is $2 \times 2 \times 2 \times 3$ (or $2^3 \times 3$).
Every composite number can be expressed with prime factors and has only one prime factorization.

Also called complete factorization.
See also factor tree, prime factor, prime number.

prime number

A prime number is a counting number that has exactly two factors, 1 and itself.

Examples

$$5 = 1 \times 5$$
$$19 = 1 \times 19$$

The number 1 has only one factor ($1 \times 1 = 1$).
Therefore, it is not a prime number.
It is also not a composite number.

See also composite number, prime factorization.

principal

Principal is an amount of money borrowed or loaned.

See also interest.

prism

A prism is a space figure with two parallel bases that are polygons of the same size and shape. All other faces of a prism are rectangles or other parallelograms. The shape of its bases gives a prism its name.

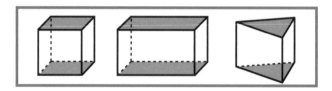

See also oblique, rectangular prism, triangular prism.

probability

Probability is the likelihood, or chance, that a given event will occur. Probability is represented with a fractional number between 0 and 1.

The closer to 1, the more likely the event is to occur.
The closer to 0, the less likely the event is to occur.

Examples

The probability that students will need to go to school on Monday mornings during nine months of the year is close to 1.
The probability of snow in Arizona during July is close to 0.
The probability that a tossed penny will land on its head is $\frac{1}{2}$.

probability
(continued)

Also called chance.

See also compound event, dependent event, independent events, outcome.

problem

A problem is a situation in which all three of the following are true:

> You want or need to find a solution.
> You do not already know how to find the solution.
> You are willing to try to find the solution.

Mathematics is learned through problem solving, which is the major reason for studying mathematics.

product

A product is the result of multiplication.

Examples

> $5 \times 6 = 30$
> $1.3 \times 6 = 7.8$
> $21 \times 11 = 231$
> factor $\times$ factor = product

See also factor, factor pair, multiplication.

profit

In business transactions, profit is the amount of money left after all expenses are paid.

Example

Pedro made some crafts to sell at the craft fair. He paid $15.43 for supplies. He sold all the crafts he made for $22.00. Not counting the time he spent making the crafts or the additional supplies he borrowed from his mother, what was Pedro's profit?

> Pedro's income was $22.00.
> Pedro's expenses were $15.43.
> $22.00 - 15.43 = 6.57$.
> Pedro's profit was $6.57.

See also loss.

proper factors

The proper factors of a number are all its factors except the number itself.

Example

The proper factors of 20 are 1, 2, 4, 5, and 10.

See also factor.

proper fraction

A proper fraction is a common fraction that names a number less than 1.

Examples

For each proper fraction, the numerator is less than the denominator.

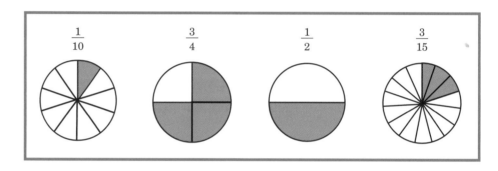

proportion

A proportion is a statement that two ratios are equal. Proportions can be written as equivalent fractions or as equal ratios.

Examples

Equivalent fraction $\frac{1}{2} = \frac{7}{14}$

Equal ratio $1:2 = 7:14$

terms of a proportion
There are four terms of a proportion.
A proportion may be written $\frac{a}{b} = \frac{c}{d}$ or $a:b = c:d$
The terms are a, b, c, and d.
The extremes are a and d, and the means are b and c.
The product of the means and the product of the extremes are equal.

Example

Proportion: $1:2 = 7:14$. Note that $2 \times 7 = 1 \times 14$.

proportion
(continued)

See also equal ratios, scale.
Related word proportional.

protractor

A protractor is a tool used
to measure angles.

See also angle.

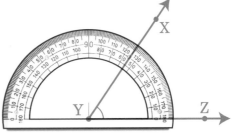

pyramid

A pyramid is a space figure that has any polygon for a base
and triangles for all other faces.

apex
The triangular faces meet at a point, or vertex, called the apex.
The shape of its base names a pyramid.

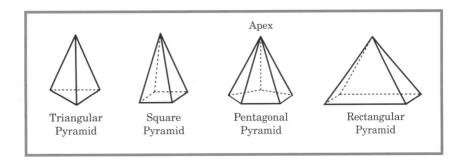

Triangular   Square   Pentagonal   Rectangular
Pyramid   Pyramid   Pyramid   Pyramid

?? Did You Know ??

The pyramids of Egypt served as tombs for ancient
Egyptian kings. Pyramids are sometimes used in
modern construction, as in the Louvre, a famous
museum in Paris (pictured below).

See also apex.

Q

quadrant

A quadrant is one of the four sections of a rectangular coordinate plane. The quadrants are separated by the *x*- and *y*-axes. The quadrants are numbered in order from I to IV, starting in the upper right quadrant and going counterclockwise.

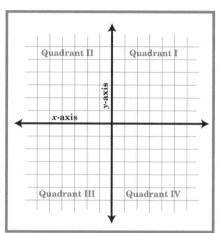

?? Did You Know ??

The first part of the word *quadrant* is from a Latin root meaning four.

A popular name for an all-terrain vehicle (ATV) is *quad*, named for its four large tires. Other words based on the same root include *quadrilateral*, *quadruplets*, and *quadrillion*.

See also coordinate system.

quadrilateral

A quadrilateral is a polygon that has four sides.

Examples

These figures are all quadrilaterals.

See also polygon.

142

quadrillion

A quadrillion is equal to 1,000 trillions. In standard form, one quadrillion is written as 1,000,000,000,000,000.

In exponential form, one quadrillion may be written as 1×10^{15} (or simply as 10^{15}).

quart (qt)

A quart is a unit of capacity in the customary system of measurement. (A quart is a little less than a liter.) A quart is usually used to measure liquids.

1 quart = 4 cups or 2 pints
4 quarts = 1 gallon

dry quart
A dry quart is used to measure dry products such as grains, berries, and other fruits. It is a little larger than the quart used in liquid measure.

1 dry quart = 2 pints
8 dry quarts = 1 peck
4 pecks = 1 bushel

See also customary system of measurement.

quarter-hour

A quarter-hour is a unit of time equal to 15 minutes.
8:15 is a quarter-hour later than 8:00.

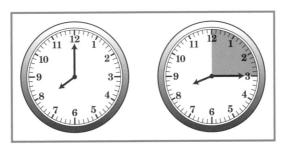

?? Did You Know ??

Notice the *quart* in *quarter*. There are four quarts in a gallon, four quarters in a dollar, four quarter-hours in an hour, and four quarters in a football game. *Quart* comes from a Latin word meaning four or fourth.

quotient

A quotient is the number resulting from division.

Examples

$$12 \div 3 = 4$$

$$4\overline{)26} \overset{6 R2}{}$$

12 is the dividend.
3 is the divisor.
4 is the quotient.

26 is the dividend.
4 is the divisor.
6 is the quotient.
2 is the remainder.

See also dividend, division, divisor, remainder.

radius

*plural **radii***

The radius of a circle is any line segment from the center of the circle to a point on the circle.

The radius of a sphere is any line segment from the center of the sphere to a point on the sphere.

Circles and spheres have an infinite number of radii.

See also circle, sphere.

Related words radiate, radii.

random

A random event happens by chance. For example, whether a tossed penny lands "heads" or "tails" is random.

range

In statistics, the range is the difference between the greatest number and the least number in a set of data.

Example

Ritchie earned the following scores in his gymnastics events:
- Floor exercise 9.4
- Rings 9.5
- Pommel horse 9.7
- Vault 9.2

The greatest number is 9.7
The least number is 9.2
To find the range, subtract the least number from the greatest.
9.7 − 9.2 = 0.5

The range in Ritchie's scores is 0.5.

See also estimation strategies, probability, rounding, statistics.

rate

A rate is a ratio that compares two quantities expressed in different kinds of units.

Examples

> 3 pens cost $1.19. Items (pens) are one kind of unit, money ($1.19) is another.

> A jet may travel at the rate of 1,000 miles per hour. Distance (miles) is one kind of unit, time (hour) is another.

> The speed of light in space is 186,000 miles per second. Distance (miles) is one kind of unit, time (second) is another.

> The rate of interest on a loan for a new car was advertised at 5.9% per year. This is a complex kind of rate. It compares a rate (percent, or parts per hundred) with another kind of unit, time (year).

See also interest, unit price.

ratio

A ratio is a comparison of two quantities using a fraction or division.

Example

> There are two people riding on this elephant. The ratio of people to elephants is $\frac{2}{1}$, or 2:1.

Notice that a ratio can be written in two ways: $\frac{a}{b}$ or $a{:}b$ (b does not equal 0). The first term of the ratio is a and the second term is b.

See also common fraction, equal ratios, proportion.

Related word rational.

rational number

A rational number is a number that can be written in the form of a common fraction.

Examples

$\frac{1}{4}, \frac{2}{3}, \frac{4}{2}, \frac{6}{1}, \frac{10}{2}$

A rational number can also be written in the form of a decimal fraction that either terminates or repeats.

Examples

0.25   terminating decimal
$0.6\bar{6}$ repeating decimal

See also common fraction, decimal fraction, fractional number, nonterminating decimal, terminating decimal.

Related word ratio.

ray

A ray is a part of a line that has one endpoint and extends infinitely in one direction.

$$\xrightarrow[\text{A} \qquad\qquad \text{B}]{\bullet\longrightarrow\quad\bullet\longrightarrow}$$

$\overrightarrow{AB}$ (ray AB)

Example

This number line showing whole numbers is really not a line in geometry, but a ray. It begins at the endpoint, 0, and extends infinitely because there is no greatest whole number.

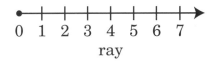

ray

ray
(continued)

R

A ray of sunshine begins at the sun (the endpoint) and travels through space in one direction. A ray from the sun can be used as a model for thinking about a ray in geometry.

See also line, line segment.

real graph

A real graph is a kind of bar graph that uses the real objects being graphed. This real graph uses baseballs.

How many hits we got in the ball game			
Terry	⊖	⊖	
Joe	⊖	⊖	⊖
Emily	⊖		
Sarah	⊖	⊖	
Max	⊖	⊖	⊖	⊖
Jason	⊖	⊖	

See also bar graph.

reciprocal

The reciprocal of a rational number is the number that, when used as a factor with the given number, will result in a product of 1.

Example

Rational Number	Reciprocal	Equation
$\frac{1}{4}$	$\frac{4}{1}$	$\frac{1}{4} \times \frac{4}{1} = 1$
$\frac{7}{8}$	$\frac{8}{7}$	$\frac{7}{8} \times \frac{8}{7} = 1$
$\frac{3}{1}$	$\frac{1}{3}$	$\frac{3}{1} \times \frac{1}{3} = 1$
2	$\frac{1}{2}$	$2 \times \frac{1}{2} = 1$

In the following equation, $\frac{b}{a}$ is the reciprocal.

$\frac{a}{b} \times \frac{b}{a} = 1$

Also called multiplicative inverse.

rectangle

A rectangle is a closed plane figure with four sides and four right angles. A rectangle with four equal sides is called a *square.*

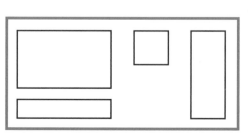

?? Did You Know ??

A rectangle in which the ratio of length to width is approximately 8 to 5, or 8:5, is called a *Golden Rectangle*. It is considered very pleasing to the eye.

The face of the Parthenon, a famous building of ancient Greece, (below, left) is a Golden Rectangle.

The United States flag is also in the shape of a Golden Rectangle.

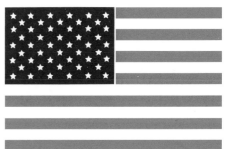

See also Golden Rectangle, parallelogram, polygon, quadrilateral, square.

Related word rectangular.

rectangular coordinate system

The rectangular coordinate system is a system used to locate points in a plane.

Also called Cartesian coordinate system.

See also Cartesian coordinate system, coordinate system.

rectangular prism

A rectangular prism is a prism that has rectangles as bases. Each of the rectangular prisms in the figures 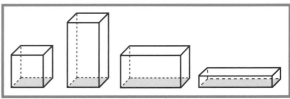 also has rectangles for its other faces. Some people call this kind of prism a right rectangular prism or cuboid. Its vertices form right angles.

Some rectangular prisms are oblique. Not all of the vertices of oblique rectangular prisms form right angles.

rectangular prism
(continued)

The word *cuboid* is formed from *cube* (a noun) and *-oid* (resembling or having the properties of the noun). Look at the drawings of the rectangular prisms, or cuboids. In what ways does each rectangular prism resemble a cube?

Also called rectangular solid.
See also base, cube, oblique, prism.

rectangular pyramid

A rectangular pyramid is a pyramid with a rectangle-shaped base.

See also pyramid.

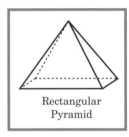

Rectangular Pyramid

reflection

A reflection is a mirror image of a figure.

Also called flip.
See also flip.

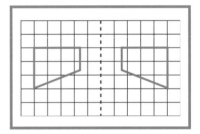

reflectional symmetry

A plane figure has reflectional symmetry if it can be divided into two congruent parts that are mirror images.

Also called line symmetry.
See also line of symmetry, line symmetry, symmetry.

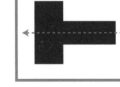

regrouping

Regrouping involves changing a number from one form to an equivalent form. In the process, the number is renamed. Regrouping is helpful in doing arithmetic computations such as the following:

regrouping
(continued)

Example

$$43$$
$$- 26$$

In long form, the regrouping is as follows (One of the tens in 43 is regrouped with the ones.):

$$
\begin{array}{rcrcr}
43 & = & 30 & + & 13 \\
- 26 & = & 20 & + & 6 \\
\hline
& & 10 & + & 7 = 17
\end{array}
$$

In short form, the regrouping is as follows:

$$
\begin{array}{r}
{\scriptstyle 1} \\
4\!\!\!/3 \\
- 26 \\
\hline
17
\end{array}
$$

?? Did You Know ??

Many years ago, the regrouping shown in the example was called *borrowing*. Regrouping is more descriptive of the action needed to solve the problem.

Also called trading.

regular polygon

A regular polygon is a plane figure with all sides of equal length and all angles of equal measure.

See also polygon.

related facts

Related facts are related addition and subtraction facts or related multiplication and division facts.

Examples

Related addition and subtraction facts using 2 and 3 as addends

$$2 + 3 = 5 \qquad\qquad 3 + 2 = 5$$
$$5 - 3 = 2 \qquad\qquad 5 - 2 = 3$$

related facts
(continued)

Related multiplication and division facts using 2 and 4 as factors

$$2 \times 4 = 8 \qquad\qquad 4 \times 2 = 8$$
$$8 \div 4 = 2 \qquad\qquad 8 \div 2 = 4$$

Also called fact family, related sentences.

See also addend, basic fact, fact family, factor.

remainder in division

In division, the remainder is the number left when one number does not divide another number exactly.

Remainders can be expressed in several ways.

remainder as a whole number
Often the remainder in division is expressed as a whole number.

Example

Floyd had 14 gumballs. He gave them to 4 friends.
How many gumballs did each person receive?

$14 \div 4 = 3$ with a remainder of 2

Each person received 3 gumballs, and there were 2 gumballs remaining. (Cutting gumballs into parts does not work very well!)

remainder as a common fraction or decimal fraction
In many division problems, the remainder is expressed as a common fraction or a decimal fraction.

Example

Jamie and Roger went on a 10-mile hike. If the trip lasted 4 hours, about how many miles did they hike per hour?

$10 \div 4 = 2\frac{1}{2}$ or 2.5

Jamie and Roger hiked about $2\frac{1}{2}$ miles (or 2.5 miles) per hour.

In this situation, the remainder is expressed as $\frac{1}{2}$ or 0.5.

remainder ignored
In some problems, the remainder is ignored.

Example

Bobbi was making bookshelves for her room.
She had 8 feet of lumber for shelves.
Each bookshelf was 3 feet long.
How many shelves could she make?

$8 \div 3 = 2$, with 2 feet of shelving remaining.

She could make 2 shelves. The remaining lumber is not enough for another bookshelf.

**remainder
in division**
(continued)

remainder as the next highest whole number
Sometimes a remainder requires the problem solution
to be the next highest whole number.

Example

Ken bought hot dog buns for the picnic.
He needed 60 buns in all.
The buns were packaged in bags containing 8 each.
How many packages of buns did he need to buy?

$60 \div 8 = 7$ with a remainder of 4

If Ken buys 7 packages of hot dog buns, he will have only 56
buns, and he needs 60. He will need to buy 8 packages of buns.

See also division, subtraction.

**remainder in
subtraction**

In subtraction, the remainder is the number left in
a "take away" situation.

Example

Karl had 15 baseball cards. He gave 9 to Jim.
How many baseball cards did he have left?

$15 - 9 = 6$

The remainder is 6, the number of baseball cards
Karl had left. (9 cards were "taken away.")

**repeated
addition**

Repeated addition is a way to multiply in which the same
number is added over and over.

Examples

$3 \times 8 = 8 + 8 + 8 = 24$
$4 \times 35 = 35 + 35 + 35 + 35 = 140$

See also multiplication.

**repeating
decimal**

A repeating decimal is a decimal number with a digit or a
group of digits that repeat on and on, without end. When
writing a repeating decimal, a bar is often placed over the
repeating portion.

repeating decimal
(continued)

Examples

$\frac{1}{3}$ expressed as a decimal is 0.3333 . . . or $0.\overline{3}$

$\frac{2}{11}$ expressed as a decimal is 0.1818 . . . or $0.\overline{18}$

See also nonterminating decimal, terminating decimal.

rhombus

plural **rhombuses** or **rhombi**

A rhombus is a parallelogram with all sides congruent.

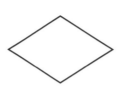 

Examples

In everyday language, some rhombuses are called diamonds—a baseball diamond, for example.

?? Did You Know ??

The term *rhomboid* can be used to name a parallelogram that has oblique angles and adjacent sides are unequal.

Related word rhomboid.

See also parallelogram, square.

right angle

A right angle is an angle that measures 90°.

See also acute angle, angle, obtuse angle, straight angle.

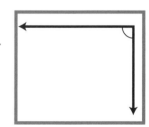

right triangle

A right triangle is a triangle with one right angle (90°).

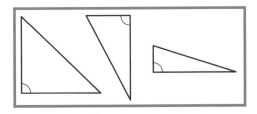

See also triangle.

rod

A rod is a unit of length in the customary system of measurement.

> 1 rod = 16.5 feet
> 160 square rods = 1 acre

Roman numeration system

The Roman numeration system probably developed between 500 BC and AD 100, during the early years of the Roman Empire. It uses Roman numbers, or letters and combinations of letters, to represent numbers. The following are Roman numerals and their values in the base-ten system:

Roman	I	V	X	L	C	D	M
Base ten	1	5	10	50	100	500	1,000

Three rules commonly observed in writing Roman numerals are:
V, L, and D can be used only once in a numeral.
I, X, C, and M can be used as many as three times in a row in a numeral.
The following "subtractions" can be performed: IV = 4, IX = 9, XL = 40, XC = 90, CD = 400, CM = 900.

The largest Roman numeral we can write following these rules is not very big: MMMCMXCIX (3,000 + 900 + 90 + 9 , or 3,999). What if you need to write a larger number? A line segment over a symbol multiplies the symbol by 1,000. For instance, $\overline{X}$ is equal to 10,000. This notation is not commonly used today because the largest numbers written in Roman numerals are usually dates.

If you visit the historic section of a city, look for Roman numerals on the cornerstones of older buildings.

See also decimal numeration system.

rotation

A rotation is the movement of a figure around a fixed point.

Also called turn.

See also turn.

rotational symmetry

A figure has rotational symmetry if, when rotated less than a full turn around a fixed point, the shape of the figure moves onto itself and looks the same.

This flower is said to have 72° rotational symmetry, or a rotational symmetry of Order 5. (When rotated, it matches itself 5 times.)

Also called point symmetry, turn.

See also symmetry, turn.

rounding

Rounding is a procedure sometimes used in estimating the result of an arithmetic operation.

rounding a whole number
Rounding a whole number usually results in a number that is close to the original number and has the same number of digits, but more of the digits are zeros.

Example

During her stay at camp, Anna had 12 hours of horseback riding lessons. She also rode on trail rides for 17 hours. About how many hours did Anna get to ride horses while she was at camp?

    12, rounded to the nearest 10, is 10.
    17, rounded to the nearest 10, is 20.
    10 + 20 = 30

Anna rode a horse about 30 hours while she was at camp.

rounding a common fraction or mixed number
One way of rounding a common fraction or common mixed number is to round to the nearest whole number.

Example

While fishing, Barbra caught, measured, and then released two sunfish. The larger one was $5\frac{3}{4}$ inches long, and the smaller one was $4\frac{1}{8}$ inches long.
About how much longer was the larger fish?

$5\frac{3}{4}$, rounded to the nearest whole number, is 6.

$4\frac{1}{8}$, rounded to the nearest whole number, is 4.

$6 - 4 = 2$

The larger fish was about 2 inches longer than the smaller one.

rounding a decimal fraction or decimal mixed number
Rounding a decimal fraction or a decimal mixed number usually results in a number close to the number being rounded, but with fewer digits.

Example

Cliff and Clara bought a large pizza for $11.25. They also bought a small pizza for $5.95. About how much did they pay for both pizzas?

$11.25, rounded to the nearest whole number (or dollar) is $11.

$5.95, rounded to the nearest whole number (or dollar) is $6.

$11 + 6 = 17$

Cliff and Clara paid about $17 for both pizzas.

See also estimation strategies.

row

A row is an arrangement of items or numbers from left to right in an array or table.

Manned U.S. Space Travel in 1960s		
Year	Number of Missions with Astronauts	Time in Space
1961		30 minutes
1962		19 hours
1963		34 hours
1964		0
1965		330 hours
1966		309 hours
1967		0
1968		407 hours
1969		873 hours

←—row (pointing to 1965 row)

See also column.

Ss

sales tax rate

Sales tax rate is tax expressed as a percent of the price the customer pays for an item. It is added to the price. In most states, the sales tax rate is set at the state level. This rate varies from state to state.

Example

Joseph bought a large chocolate chip cookie that cost $1.00. The sales tax rate in his state is 6%.
In all, how much did Joseph pay?

    6% of 1.00 is .06.
    1.00 + .06 = 1.06
Joseph paid $1.06 in all.

sample

A sample consists of all the people or items chosen to represent a larger group.

Also called sample group.
See also sampling, population.

sample space

Sample space consists of all the possible outcomes of a probability experiment.

Example

When a coin is tossed, there are two possible outcomes: heads or tails. The sample space for the experiment of tossing a coin is {heads, tails}.

See also outcome, probability.

sampling

Sampling is selecting people or items to represent a larger group. Sampling must be done carefully for the sample to be representative of the larger group, or population.

Example

Imagine you want to know the favorite movie of moviegoers in your community. If you survey only the students in your school, your sample is not likely to represent your community. Instead, you need to survey moviegoers of different age levels, different educational backgrounds, different incomes, and so on.

See also population, sample.

scale
(in measurement)

A scale is a series of marks at regular intervals along a line, used for measuring. A scale usually uses a number line.

Examples

The scales on thermometers and rulers have numbered lines.

Scale is also the name of a tool for measuring weight.

In a picture graph, the graph scale is the ratio between the picture or icon and the number it represents. In a bar graph or line graph, the graph scale is the ratio between each space on the graph and the number it represents.

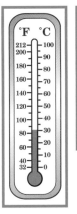

Examples

The scale for this picture graph is 1 spaceship = 6 students.

For how long would you like to travel in space?	
0 months	🚀🚀🚀
3 months	🚀🚀🚀🚀🚀🚀🚀🚀🚀🚀🚀🚀
6 months	🚀🚀🚀🚀🚀
12 months	🚀🚀🚀🚀🚀🚀🚀

Each 🚀 stands for 6 students.

scale
(in graphs)
(continued)

The scale for this bar graph is 1 square = $\frac{1}{2}$ hour.

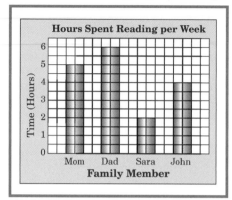

scale
(in scale drawing)

A scale drawing represents an actual object but is different in size. Scale drawings are usually smaller than the object represented. The scale for a drawing is the ratio between the size of the drawing and what is represented. A map is an example of a scale drawing.

map scale
A map scale is a ratio between the dimensions on the map and the dimensions of the area it represents.

Example
The scale for this map is located in the lower right.

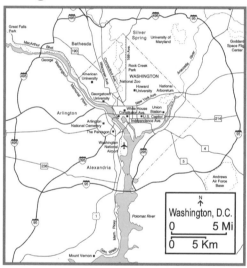

scale model

A scale model is a model of an actual object that is different in size. Usually it is smaller than the actual object. The scale for a model of an object is the ratio between the size of the model and the size of the actual object.

The photo shows a scale model of the original Stourbridge Lion, the first locomotive steam engine operated in the United States.

See also proportion, ratio, similar figures.

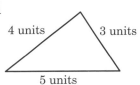

scalene triangle

A scalene triangle is a triangle with all three sides of different lengths.

See also triangle.

4 units   3 units

5 units

second

Second is the ordinal number after first. For example, second grade comes after first grade.

Second (sec) is also a unit for measuring very short lengths of time.

> 60 seconds = 1 minute
> 60 minutes = 1 hour

Other uses of second include measuring degrees of latitude and longitude.

> 60 seconds = 1 minute ($60'' = 1'$)
> 60 minutes = 1 degree ($60' = 1°$)

?? Did You Know ??

Television commercial breaks usually last about 15 or 30 seconds.

See also latitude, longitude.

semicircle

A semicircle is half of a circle. It resembles a round paper plate folded in half.

set

A set is a collection or group.
Sets may contain numbers, objects, or other items.

Examples

> Set of people in my family: Matt, Eula, Jamie, Roger.
> Set of counting numbers: 1, 2, 3, 4, ...

element
Each object of a set is called a member or element.

sharing

Sharing is a form of division in which a number is divided into equal parts.

sharing
(continued)

Also called partition division, partitive division.
See also division, partition division.

side

Side is a term that is used in both geometry and algebra.

The sides of a polygon are the three or more line segments that form the polygon.

Example

Each line segment is one side.

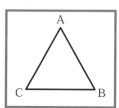

The sides of an angle are the rays that form the angle.

Example

Each ray is one side of the angle.

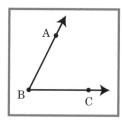

The sides of an equation are the expressions on either side of the = sign.

Example

$$4 + n = 7 - 1$$
$4 + n$ is the left side of the equation.
$7 - 1$ is the right side.

See also expression, line segment, ray.

similar figures

Similar figures have the same shape but not necessarily the same size. For polygons, corresponding angles have the same measure, and corresponding sides are proportional. All circles are similar circles.

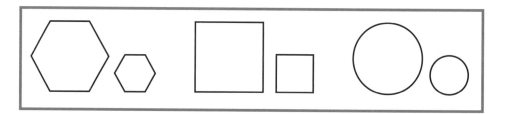

See also circle, congruent figures, proportion.

similar fractions

Similar fractions are fractions with the same denominator.

Also called like fractions.

simple closed curve	A simple closed curve is a closed curve that does not cross itself. ***See also*** closed curve.
simplest form	Both common fractions and algebraic expressions may be simplified, or written in simplest form. common fractions A common fraction is in simplest form when the greatest common factor of the numerator and the denominator is 1.

Example

To express $\frac{6}{8}$ in simplest form, think:
What is the greatest common factor (GCF) of 6 and 8?
The GCF is 2. Use 2 as the numerator and denominator of a common fraction ($\frac{2}{2}$). This common fraction is equivalent to 1.

Divide $\frac{6}{8}$ by this value for 1.

$$\frac{6}{8} \div \frac{2}{2} = \frac{3}{4}$$

Think: What is the greatest common factor of 3 and 4?
The GCF is 1.

$\frac{6}{8}$ expressed in simplest form is $\frac{3}{4}$.

?? Did You Know ??

Many years ago, expressing a fraction in simplest form was called "reducing the fraction." However, the fraction is not reduced—its value stays the same! That's why we no longer use the expression "reducing the fraction."

An expression is in simplest form when no terms can be combined.

Example

To express 3 + (2 × 5) in simplest form:
First, perform the operation within parentheses.

$2 \times 5 = 10$

Next, perform the addition.

$3 + 10 = 13$

3 + (2 × 5) expressed in simplest form is 13.

For common fractions, ***also called*** lowest terms, lowest terms of a fraction.

simplest form
(continued)

See also greatest common factor, order of operations, simplify.
Related word simplify.

simplify

To simplify is to express in simplest form.

Both common fractions and algebraic expressions may be simplified.

See also simplest form.

skew lines

Skew lines are lines in space that are not in the same plane. They do not intersect and are not parallel.

Imagine a lane on a major highway as one line and the lane or road passing over it as another line. These two lines are skew lines.

skip counting

Skip counting is counting by a given whole number greater than 1.

Examples

Skip counting by twos beginning with 2:   2, 4, 6, 8, …
Skip counting by fives beginning with 5:   5, 10, 15, …
Skip counting by tens beginning with 25:   25, 35, 45, …

Some people use skip counting as a way to figure out a multiplication fact.

Example

To figure out 6 × 7, count by 7 six times.
7, 14, 21, 28, 35, 42
6 × 7 = 42

Skip counting is also used in counting change.

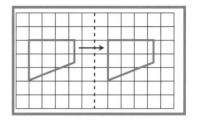

skip counting
(continued)

Example

At your garage sale, a person hands you a dollar bill for an item that costs 75 cents. Starting with 75, you skip count by 5 to 80 (one nickel) and then by 10 to 100 (two dimes). You give that person back a nickel and two dimes in change.

slide

A slide is the movement of a figure along a line.

Also called translation.
See also flip, turn.

solid

A solid is a shape that occupies space and has volume. A solid has three dimensions.

Also called solid figure, solid shape, space figure, three-dimensional figure, three-dimensional shape.
See also space figure.

space figure

A space figure is a geometric figure that occupies space and has volume. Space figures are three-dimensional.

Examples

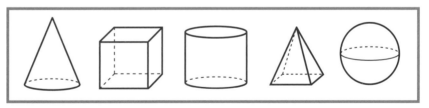

Also called solid, solid figure, solid shape, three-dimensional figure, three-dimensional shape.
See also three-dimensional.

sphere

A sphere is a space figure shaped like a round ball. All points on the sphere are an equal distance from its center.

sphere
(continued)

S

?? Did You Know ??

The earth is shaped like a sphere that has been flattened at both polar regions. This shape is called an oblate spheroid.

See also hemisphere.
Related words spherical, spheroid.

spiral

A spiral can be thought of as a curve that, as it moves farther away from its beginning point or origin, becomes less curved.

Example

Many types of shells are examples of a spiral in nature.

square

A square is a rectangle with all four sides of equal length.

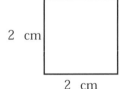

2 cm

2 cm

square of a number

The square of a number (n^2) is the result of multiplying that number by itself.

Example

6^2 may be expressed as 6×6 and is equal to 36. Some ways of reading 6^2 are "the square of 6," "six squared," and "6 to the power of 2." This drawing is a model of 6^2.

?? Did You Know ??

A square is also a rhombus. Yes, it is a parallelogram with all sides congruent! But not all rhombuses are squares.

See also exponent, power of a number, rectangle, rhombus, square number.

square centimeter (cm²)

A square centimeter is equal to the area of a square that measures 1 centimeter on each side.

See also cm², square unit.

square foot (ft²)

A square foot is equal to the area of a square that measures 1 foot on each side.

Also written as sq ft.
See also ft², square unit.

square inch (in.²)

A square inch is equal to the area of a square that measures 1 inch on each side.

Also written as sq in.
See also in.², square unit.

square kilometer (km²)

A square kilometer is equal to the area of a square that measures 1 kilometer on each side.

See also km², square unit.

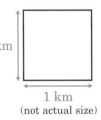
1 km
1 km
(not actual size)

square meter (m²)

A square meter is equal to the area of a square that measures 1 meter on each side.

See also m², square unit.

square mile (mi²)

A square mile is equal to the area of a square that measures 1 mile on each side.

Also written as sq mi.
See also mi², square unit.

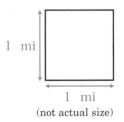
1 mi
1 mi
(not actual size)

square number

A square number is usually thought of as a counting number that can be shown as an array in the shape of a square. It is the result of multiplying some counting number by itself.

square number (continued)

Example

$$5 \times 5 = 25 \text{ or } 5^2 = 25$$

Examples

Arrays for the first five square numbers are shown.

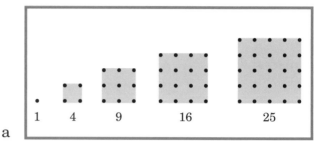

square pyramid

A square pyramid is a pyramid with a square-shaped base.

See also pyramid.

square root ($\sqrt{\ }$)

A square root of a number is one of two equal factors of that number.

Examples

The square root of 4 is 2 because
$$2 \times 2 = 4$$
The square root of 6.25 is 2.5 because
$$2.5 \times 2.5 = 6.25$$

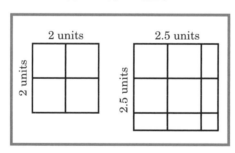

See also square number.

square unit

A square unit is equal to the area of a square with each side measuring 1 unit.

In the customary system of measurement, commonly used square units include in.^{2} (square inch), ft^2 (square foot), yd^2 (square yard), and mi^2 (square mile).

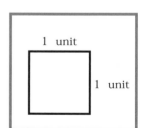

square unit
(continued)

In the metric system of measurement, commonly used square units include cm^2 (square centimeter), m^2 (square meter), and km^2 (square kilometer).

Although a square unit of area may be in the shape of a square, it can take other shapes.

Also written as sq unit, $unit^2$.
See also area, cm^2, ft^2, $in.^2$, km^2, m^2, mi^2, yd^2.

square yard
(yd^2)

A square yard is equal to the area of a square that measures 1 yard on each side.

Also written as sq yd.
See also yd^2, square unit.

standard form

Standard form is the usual way to write the name of a number, with digits. When the standard form of a whole number has more than 4 digits, it is usually written with commas separating groups of 3 digits, called periods.

Examples

34	807	2720
		(also 2,720)
17,302	1,000,000	2,103,891,762

See also expanded form, period, word form.

standard system

The standard system is the system of measurement commonly used in everyday life in the United States.

Also called customary measurement system, customary system, customary system of measurement, English measurement system, English system of measurement, standard system of measurement, U.S. Customary System.
See also customary system of measurement.

statement

A statement is a sentence that can be identified as either true or false.

statement
(continued)

Examples

$4 + 5 = 9$ (true)
$4 + 5 = 3 \times 3$ (true)
Whole numbers include counting numbers. (true)
A stop sign is in the shape of a hexagon. (false)
$4 + 5 = 7 + 1$ (false)

statistics

Statistics is the science of collecting, organizing, representing, and interpreting data. The word *statistics* is also often used to mean data.

?? Did You Know ??

A statistician uses mathematical ideas from the field of statistics to answer questions and solve problems.

This baseball card shows statistics for Babe Ruth, a famous baseball player.

Related words statistical, statistician.

BABE RUTH

Name:	George Herman Ruth
Born:	February 6, 1896
	Baltimore, MD
Died:	August 16, 1948
	New York, NY
Height:	6'2" Weight: 215 lbs.
Position:	Pitcher & Outfielder
Bats:	Left Throws: Left
Batting:	Hit 714 homers and had
	a lifetime average of .342.
Teams:	Baltimore Orioles
	Boston Red Socks
	New York Yankees
	Boston Braves

Started his career in 1914 with the Baltimore Orioles. Joined the Boston Red Sox later that year. Ended his career in 1935 with the Boston Braves.
One of the first five players elected to the National Baseball Hall of Fame.

stem-and-leaf plot

A stem-and-leaf plot is a type of graph that organizes data so that frequencies can be compared.

The stem-and-leaf plot groups the football scores shown in the table. The first column represents tens, and the second column represents ones.

1 | 1   means 11.
2 | 5, 2, 7 means 25, 22, and 27.
3 | 1   means 31.

Scores for most of the games were in the 20s.

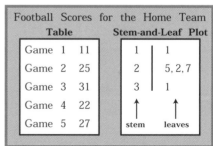

Football Scores for the Home Team

Table		Stem-and-Leaf Plot	
Game 1	11	1	1
Game 2	25	2	5, 2, 7
Game 3	31	3	1
Game 4	22	stem	leaves
Game 5	27		

straight angle

A straight angle has a measure of 180°. It is also equal to the measure of two right angles. Its measure is equal to the measure of a straight line.

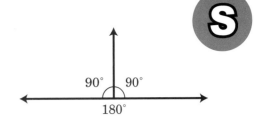

90°   90°

180°

See also angle.

straightedge

A straightedge is a tool used for drawing line segments. Many geometric figures can be constructed with only a straightedge and a compass.

An unmarked ruler is often used as a straightedge.

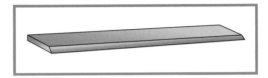

See also construction.

subtraction

Subtraction is one of the four basic arithmetic operations. (The other basic operations are addition, multiplication, and division.)

It helps us find answers to questions such as the following:

How many are left? *or* How much is left?

How many more (or fewer) are there? *or*
How much more (or less) is there?

How many more are needed? *or*
How much more is needed?

remainder

In some subtraction problems, we find the remainder, or how many are left after some are taken away.

Example

Josh raises ponies to sell.
He had 15 ponies and sold 6 of them.
How many ponies did he have left?

15 − 6 = 9

Josh had 9 ponies left (a remainder).

difference

In some subtraction problems, we need to compare two numbers to find how much more one number is than another. The answer is called a difference.

subtraction
(continued)

S

Example

David and William went tuna fishing.
David's largest catch weighed 65 pounds.
William's largest weighed 58 pounds.
How much heavier was David's largest catch?

$65 - 58 = 7$

David's largest catch weighed 7 pounds more
(the difference).

missing addend
Subtraction is one way to find a missing addend, or how many
more must be added to have the number needed.

Example

Mary has 23 eggs.
She wants to color 36 eggs for the egg hunt.
How many more eggs does she need?

$23 + ? = 36$, so $36 - 23 = 13$

The missing addend is 13, the number of eggs
Mary needs.

See also difference, missing addend, operation, remainder.

Related word subtract.

subtraction sentence

A subtraction sentence is a number sentence used to express
subtraction.

Examples

$5 - 3 = 2$
$4\frac{1}{2} - \frac{3}{4} = 3\frac{3}{4}$

See also number sentence.

sum

A sum is the result of addition.

Example

In $15 + 8 = 23$,
15 and 8 are addends.
+ is the symbol for addition.
23 is the sum.

See also addend, addition.

surface area

Surface area is the total area of the surface of a space figure.

The surface area of a rectangular prism is found by adding the areas of all the faces. To find the surface area of this gift, first find the area of each of the six faces. Then add the surface areas to find the total.

survey

A survey is usually a list of questions asked of a sample of people to determine the characteristics of a group. Surveys are often conducted in writing or by telephone. This is an example of a written survey.

> *Movie Survey:*
> 1. What is the last movie you saw? _____
> 2. How many movies do you watch each month?
>   _____
> 3. What is your all-time favorite movie?
>   _____

symmetry

In geometry, symmetry describes the balance a figure has.
Both plane and space figures may have symmetry.

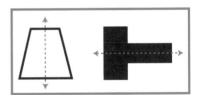

line symmetry
A plane figure has line symmetry if it can be divided into two congruent parts that are mirror images.

Notice that each line of symmetry creates a mirror image, or reflection.

plane symmetry
A space figure has plane symmetry if it can be divided into two halves that are reflections of each other.

Example

This geode and fern leaf are examples of plane symmetry in nature.

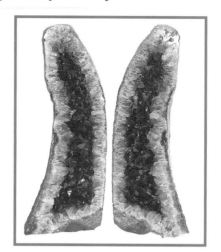

rotational symmetry

A plane figure, such as a snowflake, has rotational symmetry if, when rotated less than a full turn around a fixed point, the shape of the figure moves onto itself.

A space figure, such as a top, has rotational symmetry if the shape of the figure moves onto itself when rotated less than a full turn around a fixed point.

symmetric property of equality

Symmetry is also used to describe the symmetric property of equality.

Examples

If $3 + 4 = 7$, then $7 = 3 + 4$
If $A = l \times w$, then $l \times w = A$.

Also called flip, reflectional symmetry.

See also line symmetry, reflectional symmetry, rotational symmetry.

Related words asymmetrical, symmetrical.

table

A table is an arrangement of information in rows and columns.

Orbits of the Planets		
Planet	**Distance from Sun** (000 km)	**Orbit** (# Earth Days)
Mercury	57,910	87.97
Venus	108,200	224.70
Earth	149,600	365.26
Mars	227,940	686.98
Jupiter	778,330	4332.71
Saturn	1,429,400	10,759.50
Uranus	2,870,990	30,365.00
Neptune	4,504,300	60,190.00
Pluto	5,913,520	90,550.00

See also column, row.

tablespoon (T)

A tablespoon is a unit of capacity in the customary system of measurement.

> 1 tablespoon = 3 teaspoons
> 16 tablespoons = 1 cup

?? Did You Know ??

One or two tablespoons of peanut butter and one tablespoon of jelly make a favorite sandwich spread for many people.

See also customary system of measurement.

tally chart

A tally chart is a chart used to summarize the number of times items occur in a set of data. It is one kind of frequency table.

See also frequency table.

tally marks

Tally marks are used to record the frequency of an item. We count or keep score with tally marks.

See also frequency table.

tangram

The tangram is a Chinese mathematical puzzle made from a square cut into 7 pieces called tans. These pieces can be used to form many different geometric shapes.

Tangram is also the name of any shape made from all the tans, such as the ones pictured here.

?? Did You Know ??

According to legend, a man in ancient China named Tan dropped a porcelain tile, which broke into 7 pieces. Tan spent the rest of his life trying to put the pieces back together. As he worked with the pieces, he created more than 300 different shapes and pictures, or tangrams.

teaspoon (t)

A teaspoon is a unit of capacity in the customary system of measurement.

3 teaspoons = 1 tablespoon

See also customary system of measurement.

temperature

Temperature is a measure of heat or cold. Temperature is measured in degrees Celsius (°C) in the metric system of measurement. Temperature is measured in degrees Fahrenheit

temperature
(continued)

(°F) in the customary system of measurement.

The freezing and boiling points of water are used as reference points in both systems.

See also Celsius (°C) temperature scale, Fahrenheit (°F) temperature scale.

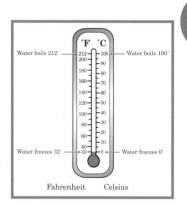

tenth

A tenth is one of 10 equal parts of a whole or group. One tenth may be written as $\frac{1}{10}$ or 0.1.

Examples

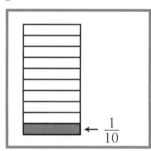 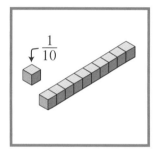

In ordinal numbers, tenth is next after ninth. For example, Natalie is the tenth person in the lunch line.

See also decimal numeration system, ordinal number.

tenths

Tenths is the name of the place to the right of the decimal point in the decimal numeration system.

In 2.98, 9 is in the tenths place.

See also decimal numeration system.

ones	decimal point	tenths	hundreths
2	.	9	8

terminating decimal

A terminating decimal is a decimal that ends. Zeroes could be placed to the right of the last digit, but the value would not change.

Examples

0.5   0.278   0.380

See also nonterminating decimal, repeating decimal.

terms

A term is a component of a fraction, ratio, proportion, or sequence.

The terms of a common fraction are its numerator and denominator.

$$\frac{3}{4}$$

← numerator
← fraction bar
← denominator

terms of a ratio

There are usually two terms of a ratio. They are called the first term and the second term.

Ratio: $\frac{3}{4}$ or 3:4
3 is the first term and 4 is the second term.

terms of a proportion

There are four terms of a proportion.
A proportion may be written $\frac{a}{b} = \frac{c}{d}$ or $a{:}b = c{:}d$.
The terms are a, b, c, and d.
Proportion: $\frac{3}{4} = \frac{6}{8}$

The terms of a sequence are the numbers in that pattern or sequence.

Example

2, 4, 8, 16, 32, 64, 128, 256, 512, 1,024, ...

The terms of an expression are the parts of an expression separated by operation signs.

Expression: $3x + 2$
$3x$ and 2 are terms.

See also denominator, numerator, pattern, proportion, ratio, simplest form.

tessellation

A tessellation is a pattern of shapes repeated to fill a plane. The shapes do not overlap and there are no gaps. This quilt is an example of a tesselation.

tessellation
(continued)

motif
The shape that is repeated, or tessellated, is called a motif.

Also called tiling.
See also pattern, tiling.
Related word tessellate.

tetrahedron

A tetrahedron is a space figure with four triangular faces. The shape is a regular tetrahedron if all the faces are congruent equilateral triangles.

Also called triangular pyramid.
See also polyhedron.

thousand

A thousand is equal to 1,000 ones, 100 tens, or 10 hundreds.

In standard form, one thousand is written as 1,000 (or 1000). With an exponent, one thousand may be written as 1×10^3 (or simply as 10^3).

See also decimal numeration system.

thousandth

A thousandth is one of 1,000 equal parts of a whole or a group. One thousandth may be written as $\frac{1}{1,000}$ or 0.001. The single unit is equal to $\frac{1}{1,000}$ of the cube.
In ordinal numbers, thousandth is next after nine hundred ninety-ninth.

See also decimal numeration system, ordinal numbers.

thousandths

Thousandths is the name of the next place to the right of hundredths in the decimal numeration system.
In 0.367, 7 is in the thousandths place.

See also decimal numeration system, thousandth.

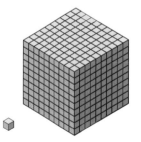

Ty Cobb's
career batting average

ones	decimal point	tenths	hundredths	thousandths
0	.	3	6	7

three-dimensional

Three-dimensional is a term used to describe space figures. They occupy space and have volume.

This box is an example of a rectangular prism, one kind of space figure. It has three dimensions: length, width, and height.

See also geometric figure, one-dimensional, two-dimensional, zero-dimensional.

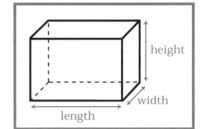

three-dimensional figure

A three-dimensional figure is a geometric figure that occupies space and has volume.

Also called solid, solid figure, solid shape, space figure, three-dimensional shape.

See also space figure.

tiling

A tiling is a pattern of shapes repeated to fill a plane. The shapes do not overlap and there are no gaps.

Also called tessellation.

See also pattern, tessellation.

timeline

A timeline is a number line that is used to show events of history in order. This timeline shows events from Einstein's life.

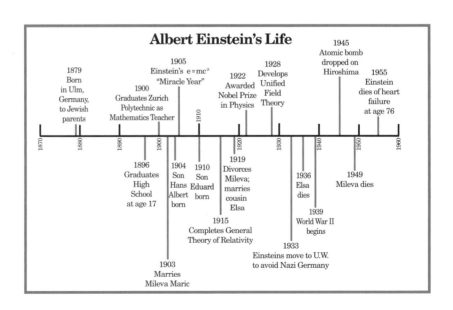

time zone

A time zone is a geographical region throughout which the same standard time is used.

This map depicts Earth's time zones.

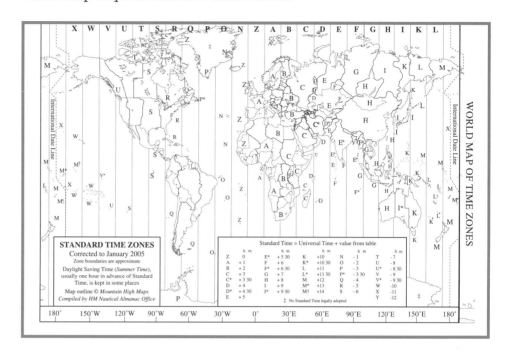

ton (T)

A ton is a unit of weight in both the metric and customary systems of measurement. In the U.S. customary system, a ton (T) is equal to 2,000 pounds.

metric ton
The metric ton (t) is equal to 1,000 kilograms (about 2,200 pounds).

?? Did You Know ??

A full-grown male African elephant may weigh about 7 metric tons, while the female may weight about half as much.

A Volkswagon "bug" weighs about 3,000 pounds, or $1\frac{1}{2}$ tons.

See also metric ton.

181

total

The total is all of a quantity or amount.

Examples

In all, 27 students were going on a field trip. Five students could ride in each car. What was the total number of cars needed?

$$27 \div 5 = 5 \text{ R2}$$
The total number of cars needed was 6.

Total can also mean "sum" or "to find the sum."
The total (sum) of 2 + 3 is 5.

Total (find the sum of) the following numbers:
$$5 + 3 + 2 + 8$$

Related word totality.

trading

Trading involves changing a number from one form to an equivalent form.

See also regrouping.

transformation

A transformation is a change in the size, shape, or position of a figure. Transformations that are changes in the position of a figure are called flips (reflections), slides (translations), and turns (rotations).

See also flip, slide, turn.

translation

A translation is the movement of a figure along a line.

Also called slide.
See also slide.

trapezoid

A trapezoid is a quadrilateral with exactly one set of parallel sides.

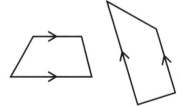

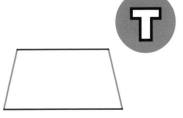

trapezoid
(continued)

isosceles trapezoid

If the nonparallel sides are the same length, the trapezoid has a special name: isosceles trapezoid. The trapezoid pictured here is an isosceles trapezoid.

?? Did You Know ??

A trapeze is in the shape of a trapezoid.

See also quadrilateral.

Related words trapeze, trapezium, trapezoidal.

tree diagram

A tree diagram is a way of organizing possible outcomes so that they are easy to count.

Example

Melissa has two kinds of ice-cream cones (sugar and waffle) and two kinds of ice cream (chocolate and vanilla).
How many possible combinations are there if she serves one scoop of ice cream on each cone?
There are four possible combinations.

	Cones	Ice Cream	Combination
	sugar cone	chocolate	sugar cone with chocolate ice cream
		vanilla	sugar cone with vanilla ice cream
ice-cream cones			
	waffle cone	chocolate	waffle cone with chocolate ice cream
		vanilla	waffle cone with vanilla ice cream

triangle

A triangle is a plane shape that has three sides and three angles. Triangles are usually named either by their sides or their angles.

These triangles are named by their sides:

equilateral triangle
All three sides of an equilateral triangle are equal in length.

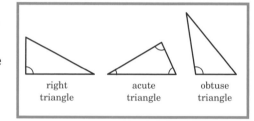

isosceles triangle
Two of the sides of an isosceles triangle are equal in length.

scalene triangle
No side of a scalene triangle is equal in length to any other side.

These triangles are named by their angles:

right triangle
One of the angles of a right triangle is a right angle (90°).

acute triangle
Each of the angles of an acute triangle is an acute angle (less than 90°).

obtuse triangle
One of the angles of an obtuse triangle is an obtuse angle (greater than 90°).

?? Did You Know ??

The *tri-* in triangle means three. Related words are *tricycle*, *triathlon*, *tripod*, and *triplet*. A triangle is often used in construction, such as the bridge at right, because it is a very strong shape.

See also polygon.

Related words triangular, triangulate

triangular number

A triangular number is a number that can be shown in an array that looks like a triangle.

Examples

Arrays for the first five triangular numbers are shown.

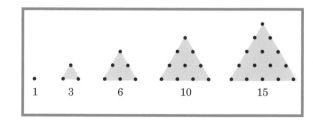

| 1 | 3 | 6 | 10 | 15 |

triangular prism

A triangular prism is a prism that has triangles as bases.

See also prism.

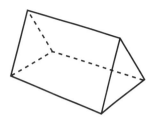

triangular pyramid

A triangular pyramid is a pyramid with a triangle-shaped base.

If all the faces are equilateral triangles, it is known as a regular tetrahedron.

Also called tetrahedron.
See also pyramid, tetrahedron.

trillion

A trillion is equal to 1,000 billions. In standard form, one trillion is written as 1,000,000,000,000. With an exponent, one trillion may be written as 1×10^{12} (or simply as 10^{12}).

See also decimal numeration system.

turn

A turn is the movement of a figure about a fixed point.

Also called rotation.
See also slide, flip.

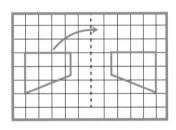

twenty-four-hour clock

A twenty-four-hour clock represents time beginning and ending at midnight.

Some times on the twenty-four-hour clock and their twelve-hour-clock equivalents are:

twenty-four-hour clock	twelve-hour clock
0100	1:00 AM
0400	4:00 AM
1300	1:00 PM
1545	3:45 PM

two-dimensional

Two-dimensional is a term used to describe plane figures.

See also geometric figure, one-dimensional, three-dimensional, zero-dimensional.

two-dimensional figure

A two-dimensional figure is a geometric figure that has area but no thickness. It lies entirely in one plane.

A rectangle is one example of a two-dimensional figure. Its dimensions are length and width.

Also called plane figure, plane shape, and two-dimensional shape.

See also one-dimensional, plane figure, three-dimensional, zero-dimensional.

unit price

Unit price is the cost per item or cost per unit of measure. Unit price can be expressed as a ratio.

Example: cost per item

Garth bought 2 erasers for $0.50. How much did each eraser cost?

> 2:50 = 1:25
> Each eraser cost $0.25.
> The unit price is $0.25.

Example: cost per unit of measure

A 1-pound loaf of bread cost $1.60. What was the cost of the bread per 1-ounce serving? (1 pound = 16 ounces)

> 1.60:16 = 0.10:1
> The unit price is $.10.

Also called cost per unit and unit cost.

See also equal ratios.

unit fraction

A unit fraction is a fraction with a numerator of 1. A unit fraction can be expressed as the sum of two equal unit fractions.

Examples

> $\frac{1}{2} = \frac{1}{4} + \frac{1}{4}$
> $\frac{1}{3} = \frac{1}{6} + \frac{1}{6}$

A unit fraction can also be expressed as the sum of two different unit fractions.

Examples

> $\frac{1}{2} = \frac{1}{3} + \frac{1}{6}$
> $\frac{1}{3} = \frac{1}{4} + \frac{1}{12}$

unit fraction
(continued)

Other fractions can also be named as the sum of unit fractions.

Examples

$$\frac{2}{3} = \frac{1}{2} + \frac{1}{6}$$

$$\frac{2}{7} = \frac{1}{4} + \frac{1}{28}$$

?? Did You Know ??

An archaeological discovery called the Rhind papyrus shows that Egyptians used unit fractions in their mathematical work well over 3,000 years ago. The Rhind papyrus is about 18 feet long and 1 foot wide. It is in the British Museum in London, England.

unlike fractions

Unlike fractions are fractions with different denominators.

Example

$\frac{1}{3}$ and $\frac{1}{2}$ are unlike fractions.
The denominator of $\frac{1}{3}$ is 3.
The denominator of $\frac{1}{2}$ is 2.

See also like fractions.

U.S. Customary System

The U.S. Customary System is the measurement system used most commonly in the United States.

Also called customary measurement system, customary system, customary system of measurement, English measurement system, English system of measurement, standard system of measurement.

See also customary system of measurement.

variable

In algebra, a variable is usually a letter or other symbol that stands for a number or quantity. Variables are used to represent several mathematical ideas. Some common uses are:

To represent unknowns that do not vary

Example

$$x + 5 = 8$$
$$13 - ? = 4$$

To represent quantities that vary

Example

The area of any rectangle can be represented by the formula $A = l \times w$.

The area (A) varies as either the length (l) or width (w) of the rectangle varies.

To represent properties

Example

The commutative property of addition can be represented by the equation $a + b = b + a$.

For any addends (variables a and b), the order in which they are added does not affect the sum.

$$3 + 4 = 4 + 3$$
$$1\tfrac{1}{4} + 7\tfrac{3}{8} = 7\tfrac{3}{8} + 1\tfrac{1}{4}$$

See also algebra.

Venn diagram

A Venn diagram is a visual way to show relationships among sets. It is usually made with circles.

Venn diagram A shows the sets of factors of 12 and 18. The factors that are the same for both numbers are included in the overlapping region. These members, or elements, are the same for the two sets.

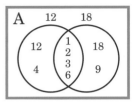

Venn diagram B shows that all squares are rectangles. Therefore, squares are a subset of rectangles. It also shows that there are rectangles that are not squares.

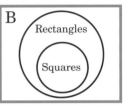

disjoint sets

Venn diagram C shows that the sets of odd numbers and even numbers do not share any elements. (A number cannot be both odd and even.) Since the sets of odd numbers and even numbers do not share any elements, they are called disjoint sets.

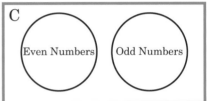

?? Did You Know ??

Venn diagrams are named for John Venn, an Englishman who lived from 1834 to 1923. Venn studied logic and developed diagrams as tools for logical thinking.

vertex

plural **vertices**

A vertex is a point at which two or more sides or edges of a geometric figure meet.

Examples

Point A is the vertex of this angle.

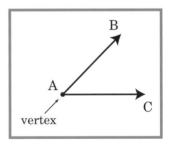

Points A, B, C, and D are the vertices of this rectangle.

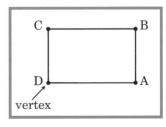

vertex
(continued)

Points A, B, C, D, and E are the vertices of this square pyramid.

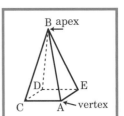

apex
Point B is the vertex at the highest point in relation to the base of this pyramid. The highest point is known as the apex.

See also angle, apex.
Related word vertices.

vertical bar graph

In a vertical bar graph, the bars, or rectangles representing the data, run from the bottom toward the top of the graph.

See also bar graph, horizontal bar graph.

vertical axis

The vertical axis is the vertical number line in a rectangular coordinate system.

Also called *y*-axis.
See also coordinate system, *y*-axis.

volume

The volume of a space figure is how much space it occupies. The volume of a container is how much it can hold.

Volume of space figures is often measured in cubic units.

cubic unit
Each edge of each square face of a cubic unit measures 1 unit.

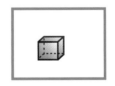

Examples

The volume of this rectangular prism is 36 cubic units.

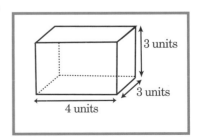

3 units

3 units

4 units

191

volume
(continued)

Below are some space figures and the formulas for finding their volume.

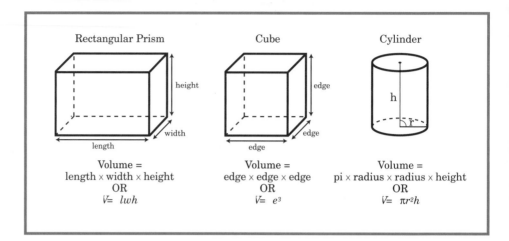

Rectangular Prism

height

width

length

Volume =
length × width × height
OR
$V = lwh$

Cube

edge

edge

edge

Volume =
edge × edge × edge
OR
$V = e^3$

Cylinder

h

r

Volume =
pi × radius × radius × height
OR
$V = \pi r^2 h$

In the metric system of measure, units of cubic measure include cm^3, dm^3, and m^3. In the customary system, units of cubic measure include in.^{3}, ft^3, and yd^3.

When a container is used for holding liquids, its volume may be measured in units of capacity (cups, quarts, liters, and so on).

See also capacity, cubic unit.

weight

Weight is how heavy an object is.

One way to find the weight of an object is to multiply the mass of the object by the gravity. (The abbreviation for *gravity*, g, is the same as the abbreviation for gram. Be careful not to confuse these two uses.)

Example

At sea level on Earth, gravity is 1, or 1 g.
The moon's gravity is $\frac{1}{6}$ of the gravity on Earth, or $\frac{1}{6}$ g.

When *Apollo 15* was launched from Cape Kennedy, Florida, in July 1971, it carried the first Lunar Rover. This vehicle, which looked a lot like a dune buggy, weighed about 462 pounds on Earth. It weighed about 77 pounds when it landed on the moon.

Although weight is not the same as mass, people often use them to mean the same outside the field of science.

?? Did You Know ??

You would weigh slightly less at the top of Mount Everest than you would at sea level because the gravity at the summit is a little less than 1 g.

See also customary system of measurement, metric system of measurement.

Related word weigh.

193

whole numbers

Whole numbers are the numbers 0, 1, 2, 3, 4, ... They include all the counting numbers and 0.

A number line is a line (or line segment or ray) on which numbers are assigned points. It helps us see numbers in relation to each other. The arrow shows that they go on without end.

On this number line showing whole numbers, we can see that 3 < 4 because 3 is closer to 0.

See also counting numbers.

width

Width is usually thought of as the measure of an object from side to side.

The arrows indicate the width of this door.

See also length.

withdrawal

A withdrawal is money removed from a checking account or savings account.

See also deposit.
Related word withdraw.

word form

The word form of a number is the number written as it is said in words.

Examples

The word form for 14 is fourteen.
The word form for $\frac{1}{2}$ is one half.
The word form for 5.6 is five and six tenths.
(Notice that the decimal point is read as *and*.)
The word form for 2,010 is two thousand ten.
(Notice that the word *and* is not included in the word form.)
The word form for 1,000,000 is one million.

Also called word name for a number.
See also expanded form, standard form.

**x- and
y-axes**

The *x*- and *y*-axes are the number lines used as references in a rectangular coordinate system.

See also coordinate system, *x*-axis, *y*-axis.

x-axis

The *x*-axis is the horizontal number line in a rectangular coordinate system.

Also called horizontal axis.
See also coordinate system, horizontal axis.

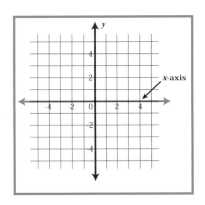

**x-and y-
coordinates**

The *x*- and *y*-coordinates are the two numbers in an ordered pair, used to identify a location on a map or in a rectangular coordinate system.

See also coordinate system, *x*-coordinate, *y*-coordinate.

x-coordinate

The *x*-coordinate indicates a distance along the *x*-axis, or horizontal axis, in a rectangular coordinate system. It is the first of two numbers in an ordered pair.

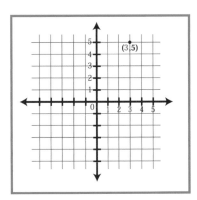

Example

In the ordered pair (3, 5)
3 is the *x*-coordinate.

See also coordinate system, ordered pair, *y*-coordinate.

195

yard (yd)

A yard is a unit of length in the customary system of measurement.

> 1 yard = 3 feet or 36 inches

?? Did You Know ??

In ancient times a yard may have been the length of the sash, or gird, worn around the king's waist.

King Henry I of England is believed to have decreed a yard to be the distance from the tip of his nose to the end of his thumb.

See also customary system of measurement.

y-axis

The y-axis is the vertical number line in a rectangular coordinate system.

Also called vertical axis.
See also coordinate system.

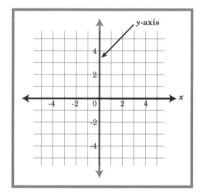

y-coordinate

The y-coordinate indicates a distance along the y-axis, or vertical axis, in a rectangular coordinate system. It is the second number in a set of two numbers called an ordered pair.

Example

> In the ordered pair (3,5) 5 is the y-coordinate.

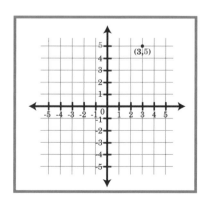

See also coordinate system, ordered pair, x-coordinate.

yd²

Read as *square yard*.

A yd² is equal to the area of a square that measures 1 yard by 1 yard.

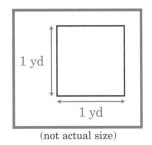

1 yd

1 yd

(not actual size)

?? Did You Know ??

A big-screen television may have a screen with an area of more than 1 square yard.

Also written as square yard, sq yd.
See also square unit, square yard.

yd³

Read as *cubic yard*.

A yd³ is equal to the volume of a cube that measures 1 yard on each edge.

Also written as cubic yard.
See also cubic unit, volume.

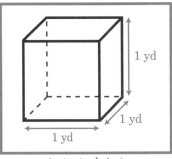

1 yd

1 yd

1 yd

(not actual size)

Zz

zero

Zero is represented by the symbol 0 in our system of numeration. It has many uses. Some of these uses are given below.

zero as a whole number

Zero is the first whole number in the set of whole numbers. (In the set of integers, zero is neither positive nor negative.)

empty set

Zero is the number of members in an empty set.

Example

The number of basketball players taller than 9 feet is 0. There are no members in this set.

placeholder

In a number with more than one digit, 0 is the placeholder (holds the place) when there is no value for that place.

Examples

In the number 40, zero holds the ones place because there are no ones. In the number 1008, zero holds the hundreds and tens places because there are no hundreds or tens.

identity for addition

Zero is the identity for addition. That is, the sum of 0 and any addend (number to be added) is equal to that addend.

Examples

$$0 + 3 = 3$$
$$\frac{1}{4} + 0 = \frac{1}{4}$$

zero property of multiplication

The zero property of multiplication means that the product of any number and 0 is equal to 0.

Examples

$$5 \times 0 = 0$$
$$0 \times 1,000,000 = 0$$

zero
(continued)

zero as an exponent
When n is not equal to zero, the value of n^0 is 1.

Example

$10^2 = 100$, $10^1 = 10$, and $10^0 = 1$.
Therefore, $258 = (2 \times 10^2) + (5 \times 10^1) + (8 \times 10^0)$,
or $200 + 50 + 8$.

?? Did You Know ??

If your great-great-grandparents were alive today and helping you with your mathematics homework, they might read the number 400 as "4 double nought." Nought (sometimes spelled *naught*) was used for a long time as the word for zero in English-speaking countries. (The symbol for nought is pictured here.) How might this meaning be related to other meanings of the word *naught*?

See also empty set, identity property for addition, zero property of multiplication.

zero-dimensional

A point is zero-dimensional because it does not occupy space. We cannot draw a picture of a point, since it has zero dimensions. We usually represent a point with a dot.

See also one-dimensional, three-dimensional, two-dimensional.

zero property of multiplication

The zero property of multiplication means that the product of any number and 0 is equal to 0.

Examples

5×0 (or 0×5) $= 0$
$\frac{3}{4} \times 0$ (or $0 \times \frac{3}{4}$) $= 0$
$^-2 \times 0$ (or $0 \times {}^-2$) $= 0$
$1{,}000{,}000 \times 0$ (or $0 \times 1{,}000{,}000$) $= 0$

The zero property of multiplication is written with symbols as $a \times 0 = 0 \times a = 0$

Also called multiplication property of zero.

3-D Earth Icosahedron

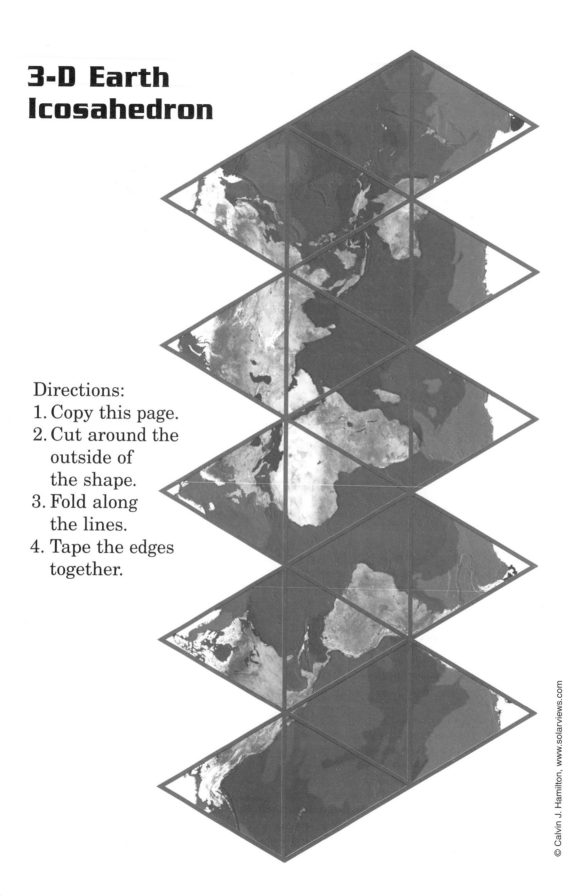

Directions:
1. Copy this page.
2. Cut around the outside of the shape.
3. Fold along the lines.
4. Tape the edges together.

Symbols

+	plus, as in 3 + 4	×	times, as in 5 × 7
−	minus, as in 6 − 2	÷	divided by, as in 9 ÷ 3
=	is equal to	≠	is not equal to
>	is greater than	<	is less than
≥	is greater than or equal to	≤	is less than or equal to
$^+4$	positive 4	$^-4$	negative 4
\$	dollar sign	¢	cent sign
%	percent	π	pi (approximately 3.14)
°	degree	…	continuing without end
°C	degrees Celsius	$1.\bar{3}$	repeating decimal 1.33 …
°F	degrees Fahrenheit	$\sqrt{\ }$	square root
′	foot, ft	″	inch, in.
′	minute, min	″	second, sec
∅	nought (zero)	∞	infinity
$\overleftrightarrow{AB}$	line AB	$\overrightarrow{AB}$	ray AB
$\overline{AB}$	line segment AB	∠ABC	angle ABC
(3, 4)	ordered pair, 3, 4	(x, y)	point coordinates on a plane
卌 IIII	tally marks	#	number, or pound, lb

( ) parentheses, shows order of operations
For example, 2 + (3 × 5) = 2 + 15 = 17

: is to, as in proportions, a:b as c:d, which means a/b = c/d
For example, 2:5 represents the ratio of 2 to 5

Table of Measures

• METRIC •

Length

1 millimeter (mm)	=	0.001 meter (m)
1 centimeter (cm)	=	0.01 meter
1 decimeter (dm)	=	0.1 meter
1 dekameter (dam)	=	10 meters
1 hectometer (hm)	=	100 meters
1 kilometer (km)	=	1,000 meters

Mass/Weight

1 milligram (mg)	=	0.001 gram (g)
1 centigram (cg)	=	0.01 gram
1 decigram (dg)	=	0.1 gram
1 dekagram (dag)	=	10 grams
1 hectogram (hg)	=	100 grams
1 kilogram (kg)	=	1,000 grams
1 metric ton (t)	=	1,000 kilograms

Capacity

1 milliliter (mL)	=	0.001 liter (L)
1 centiliter (cL)	=	0.01 liter
1 deciliter (dL)	=	0.1 liter
1 dekaliter (daL)	=	10 liters
1 hectoliter (hL)	=	100 liters
1 kiloliter (kL)	=	1,000 liters

In the metric system there is also a relationship between the units of capacity and the cubic units:

1 liter	=	1,000 cubic centimeters (cm^3)
1 milliliter	=	1 cubic centimeter

Volume

1 cubic centimeter (cm^3)	=	1,000 cubic millimeters (mm^3)
1 cubic decimeter (dm^3)	=	1,000 cubic centimeters (cm^3)
1 cubic meter (m^3)	=	1,000,000 cubic centimeters (cm^3)

Area

1 square centimeter (cm^2)	=	100 square millimeters (mm^2)
1 square meter (m^2)	=	10,000 square centimeters (cm^2)
1 hectare (ha)	=	10,000 square meters (m^2)
1 square kilometer (km^2)	=	1,000,000 square meters (m^2)

Table of Measures

• CUSTOMARY •

Length

1 foot (ft)	=	12 inches (in.)
1 yard (yd)	=	36 inches
		3 feet
1 mile (mi)	=	5,280 feet
		1,760 yards
		320 rods
1 rod	=	16.5 feet
		5.5 yards

Mass/Weight

1 pound (lb)	=	16 ounces (oz)
1 ton (T)	=	2,000 pounds

Capacity

1 cup (C)	=	8 fluid ounces (fl oz)
1 pint (pt)	=	2 cups
1 quart (qt)	=	2 pints
1 gallon (gal)	=	4 quarts
1 peck (pk)	=	8 quarts
		2 gallons
1 bushel (bu)	=	4 pecks

Area

1 square foot (ft^2)	=	144 square inches ($in.^2$)
1 square yard (yd^2)	=	9 square feet
1 acre	=	43,560 square feet
		4,840 square yards
1 square mile (mi^2)	=	640 acres

Time

1 minute (min)	=	60 seconds (sec)
1 hour (hr)	=	60 minutes
1 day (d)	=	24 hours
1 week (wk)	=	7 days
1 month (mo)	=	28, 29, 30, or 31 days
1 year (yr)	=	12 months
		52 weeks
		365 (or 366) days
1 decade	=	10 years
1 century (c)	=	100 years
1 millennium	=	1,000 years

Volume

1 cubic foot (ft^3)	=	1,728 cubic inches ($in.^3$)
		7.5 gallons
		0.8 bushel
1 cubic yard (yd^3)	=	27 cubic feet
1 gallon (gal)	=	231 cubic inches

Photograph Acknowledgments

Cover, p. 44 © Pacific Cycle, Inc., www.pacific-cycle.com

pp. 1, 26 (ice-cream cones), 27, 41, 76, 82, 92, 103, 136, 174 (fern & top) © Jupitermedia. All rights reserved. Reprinted with permission.

pp. 7, 39, 141 (protractor) courtesy Dick Blick Art Materials

pp. 9, 156 © Eula Ewing Monroe

pp. 22, 47, 51 (cups), 55, 58, 69, 79, 95, 106 (cookies, ice-cube tray), 113, 173, 194 Ralph Liberto

p. 35 © William Weatherstone, www.thedieselgypsy.com

p. 48 Photo courtesy of GeekPhilosopher.com—your premier source for free stock photos.

pp. 80, 200 © Calvin J. Hamilton, www.solarviews.com

p. 89 Premier Kites & Designs, www.premierkites.com

p. 127 © Space Imaging

p. 133 The "Human Sundial" concept by Modern Sunclocks. For details, visit their Web site at www.sunclocks.com.

p. 146 © Georgia B. and Howard O. Davis

p. 160 © Wayne County Historical Society

p. 174 Pisces Trading Company, www.bestcrystals.com

p. 181 Produced by HM Nautical Almanac Office, © Council for the Central Laboratory of the Research Councils

References

Carroll, J. B. (Ed.). *Language, Thought, and Reality: Selected Writings of Benjamin Lee Whorf.* Cambridge: Technology Press of Massachusetts Institute of Technology, 1956.

National Council of Teachers of Mathematics. *Assessment Standards for School Mathematics.* Reston, VA, 1995.

National Council of Teachers of Mathematics. *Curriculum and Evaluation Standards for School Mathematics.* Reston, VA, 1989.

National Council of Teachers of Mathematics. *Principles and Standards for School Mathematics.* Reston, VA, 2000.

National Council of Teachers of Mathematics. *Professional Standards for Teaching Mathematics.* Reston, VA, 1991.

Reutzel, D. R. and R. B. Cooter. *Teaching Children to Read: Putting the Pieces Together.* 4th ed. Upper Saddle River, NJ: Merrill/Prentice Hall, 2004.

About the Author

Formerly a classroom teacher in her native Kentucky and now a teacher educator at Brigham Young University (BYU) in Provo, Utah, Dr. Monroe has devoted her career to education at virtually all levels. Active with staff development in public schools, her passion for mathematics is a driving force in her mathematics education courses at BYU.

An author, speaker, and consultant, she is especially interested in the intimate link between language and mathematics and enjoys helping teachers and their students understand and use the language of mathematics. She claims that "math is all about relationships"—it's no wonder she sees the wonder and interconnectedness of nature through the eyes and mind of a mathematician during her world travels. But she is always eager to return home to the people and values that anchor her life—her family, friends, and spiritual community.